The Shape of All Things

A Journey Through the Map of Mathematics

Joel Peña Muñoz Jr

Dedicated to all the teachers who struggled with themselves to learn.

To my family—thank you for letting me wander off into the abyss of knowledge.

To the adults who stepped into my life when I had no father.

To my friends, who put up with my wild ideas as I learned how to think.

And to the future cyborgs, emerging. I was already talking to you.

"We did not invent the map.
We merely learned to read it."

— *The Shape of All Things*

Abstract

This book is a guided journey through the landscape of mathematics, from its ancient origins to its modern frontiers. Structured as a conceptual atlas, it illuminates how mathematical thought evolves—through number, shape, structure, logic, and uncertainty. Each chapter explores a different domain, revealing not just its content, but its character: the questions it asks, the patterns it uncovers, and the forms of reasoning it demands.
Rather than presenting mathematics as a collection of facts or procedures, this work frames it as an evolving language of understanding—a map, a mirror, and a flame. Designed for thinkers, teachers, and learners alike, it invites the reader to see mathematics not as something separate from life, but as its deep structure. The book blends narrative clarity with conceptual precision, making it accessible without compromise.
It is both a reference and a revelation. And it ends not with answers, but with new worlds to explore.

Contents

Abstract **1**

Preface **17**

Prologue – The Language of the Cosmos **19**

Reading the Map 21

Begin Where Pattern Waits 21

1 Logic, Sets, and the Skeleton of Thought **23**

1.1 What Is a Truth? 24

1.2 Propositions, Proofs, and Paradoxes 25

1.3 Sets as Spatial Logic 26

1.4 From the Empty Set to Infinity 27

1.5 Structure Before Symbols 29

1.6 The Silent Architecture of Everything 31

2 The Numbers That Built Reality **33**

2.1 The Natural Numbers — Counting What Exists 34

2.2 The Integers — Accounting for Loss 35
2.3 The Reals — Filling the Gaps 38
2.4 The Complex — Completing the System . 40

3 Algebra – The Shape of Rules 45

3.1 Symbols That Obey 46
3.2 Equations as Equal Structure 49
3.3 Variables and Generalization 52
3.4 Operations and Inverses 56
3.5 Symmetry, Groups, and the Hidden Geometry of Rules 60
Summary – From Symbol to Structure 64

4 Geometry – Space Made Visible 65

4.1 The World as Shape 66
4.2 Lines, Angles, and the Grammar of Space 68
4.3 Triangles, Proof, and the First Theorems . 73
4.4 Circles, Arcs, and the Logic of Rotation . . 76
4.5 Proof Without Numbers: Geometry as Certainty Itself 80

5 Calculus – The Mathematics of Change 83

5.1 The Curve That Refused to Be Flat 84
5.2 The Derivative: Zooming Into Change 87

5.3 The Integral:
Slicing into Wholeness 90
5.4 Limits, Infinity, and the Edge of Definition 93
5.5 The World That Can Now Be Measured . 96

6 Probability – The Logic of Uncertainty 99

6.1 The Coin That Never Promised You Anything 100
6.2 Bayes' Rule: Updating What You Believe . 103
6.3 Entropy and the Measure of Surprise . . . 106
6.4 Randomness, Pattern, and the Shape of Noise 109
6.5 Probability as Strategy: Betting, Decisions, and Belief 113
6.6 Limits, Infinity, and the Edge of Definition 117

7 Geometry – The Logic of Space 121

7.1 From Land and Shadow: The Birth of Geometry 122
7.2 Euclidean Order: Proof, Axioms, and the Infinite Flat Plane 125
7.3 The Crisis of the Fifth Postulate: Birth of Non-Euclidean Worlds 128
7.4 Curved Spaces, Bent Realities: Geometry After Euclid 130
7.5 Topology: When Shape Ignores Distance 133

8 **Logic – The Shape of Truth** **135**
8.1 The Axiom and the Argument 136
8.2 Truth Tables and Boolean Bones 139
8.3 Proving with Symbols: The Rise of Formal Logic 142
8.4 Gödel's Crack in the Foundation 145
8.5 Logic Beyond Logic: Paradox, Incompleteness, and Infinity 147

9 **Computation – The Engine of Formal Thought** **149**
9.1 The Algorithm as Essence 150
9.2 Turing's Universal Machine 152
9.3 Halting, Oracle, Limit 154
9.4 Complexity: Not Everything Easy Is Fast . 156
9.5 Computation Beyond Silicon 157

10 **Information – The Fabric of Communication** **159**
10.1 From Meaning to Measure 160
10.2 Shannon's Entropy: Information as Surprise 161
10.3 Compression and the Limits of Knowledge 163
10.4 Signal, Noise, and Channel Capacity 165
10.5 Information as Fundamental: Physics, Biology, Mind 167

11 Networks – The Shape of Connection 169

11.1 Centrality and Influence – Who Matters, and Why 173

11.2 The Small-World Principle – Closeness in Vastness 175

11.3 Emergence and Phase Shifts – When Structure Wakes Up 177

11.4 The Networked Self – Identity as Position 178

12 Category – The Mathematics of Meaning 181

12.1 Beyond Objects – Arrows as First-Class Citizens 182

12.2 Functors – Mapping Structures Between Worlds 185

12.3 Natural Transformations – Harmony Across Layers 187

12.4 Universal Properties – Essence Without Content 189

12.5 Category as Philosophy – Mathematics Reflecting on Itself 190

13 Topology – The Mathematics of Shape Without Distance **193**

13.1 Topological Spaces – When Stretching Is Allowed 194

13.2 Homeomorphism – Identity Through Deformation 197

13.3 Open Sets and Neighborhoods – The Breath of Space 199

13.4 Connectedness, Compactness, and Continuity 202

13.5 Applications – From Knots to the Cosmos 204

14 Algebra – The Language of Symmetry **207**

14.1 Algebraic Structures – Operations That Behave 208

14.2 Groups – The Anatomy of Symmetry . . . 210

14.3 Rings and Fields – Number Systems Reimagined 212

14.4 Homomorphisms – Maps That Preserve Meaning 214

14.5 The Reach of Algebra – From Crystals to Codes 216

15 Linear Algebra **219**

15.1 Vectors – Quantities with Direction and Magnitude 220

15.2 Matrices – Machines That Move Space . . 222

15.3 Linear Systems – When Equations Become Geometry 224

15.4 Eigenvectors and Eigenvalues – The Invariant Directions 226
15.5 The Power of Linearity – Why It Works Everywhere 228

16 Statistics – Patterns in the Noise **231**
16.1 What Is Statistics? From Description to Decision 232
16.2 Descriptive Statistics – What the Data Shows 234
16.3 Inference – Learning from Samples 237
16.4 Regression – Modeling Relationships 239
16.5 Bayesian Thinking – Learning as Updating Beliefs 241

17 Combinatorics – The Art of Counting **245**
17.1 What Is Combinatorics? From Counting to Configuration 246
17.2 Permutations and Factorials – When Order Matters 249
17.3 Combinations – When Order Doesn't Matter 250
17.4 Inclusion-Exclusion – Correcting for Overlaps 252
17.5 Pigeonhole Principle – Guarantees from Scarcity 254

18 Set Theory – The Foundation of Mathematics **257**

18.1 What Is a Set? Defining by Belonging 258

18.2 Operations on Sets – Union, Intersection, and Beyond 261

18.3 Subsets and Power Sets – The Structure of Possibilities 263

18.4 Infinite Sets – Beyond Counting 265

18.5 Foundations and Paradoxes – Where Definitions Collapse 267

19 Mathematical Logic **269**

19.1 Propositions and Truth Values – The Atoms of Logic 270

19.2 Logical Equivalence and Proof – Transforming Truth 272

19.3 Predicate Logic – Describing the World with Variables . . . 275

19.4 Incompleteness and Decidability – The Limits of Logic 277

19.5 Logic and Computation 279

20 Category Theory – The Mathematics of Mathematics **281**

20.1 What Is a Category? Objects, Arrows, and Composition 282
20.2 Functors – Maps Between Categories 284
20.3 Natural Transformations – Comparing Functors 286
20.4 Universal Properties – Defining by Uniqueness 288
20.5 Diagrams and Commutativity – Reasoning with Arrows 290
20.6 Category Theory as a Unifying Language 292

21 Infinity – Numbers Without Bound **295**
21.1 Potential and Actual Infinity 296
21.2 Cardinality – Comparing Sizes of the Infinite 297
21.3 Ordinal Numbers – The Order of the Infinite 299
21.4 The Continuum Hypothesis – Gaps in the Infinite 301
21.5 Infinity in Calculus, Physics, and the Real World 303

22 Chaos and Fractals – Order in Disorder **307**
22.1 Sensitivity to Initial Conditions – The Butterfly Effect 308

22.2 Strange Attractors and Geometric Complexity 311
22.3 Fractals – Infinite Detail from Simple Rules 313
22.4 Chaos in Nature – Weather, Biology, and the Brain 315
22.5 The Edge of Predictability – Chaos as a Creative Force 317

23 Mathematical Physics – The Code Beneath Reality **319**
23.1 Mathematics as the Language of Nature . 320
23.2 The Structure of Space and Time 321
23.3 Quantum Mathematics – Probability at the Core 322
23.4 Symmetry and Conservation – Noether's Theorem 325
23.5 The Unreasonable Effectiveness of Mathematics 327

24 The Mathematics of Intelligence — Learning Machines **329**
24.1 Probability as Prediction 330
24.2 Representing Data in High-Dimensional Spaces 333
24.3 Neural Networks and Transformers 346
24.4 Optimization and Learning 357

24.5 Scaling Laws — The Mathematics of Bigger, Better, and Beyond 371

25 Mathematical Beauty – Why Truth Feels Elegant 377

25.1 Elegance as Compression – Simplicity with Power 378

25.2 Symmetry and Surprise – The Dual Nature of Elegance 379

25.3 Aesthetic Selection – Why Mathematicians Trust Beauty 380

25.4 When Beauty Misleads – The Limits of Elegance 383

25.5 Beauty as Insight – A Human Need for Coherence 385

26 Mathematical Creativity – Invention Within Constraint 387

26.1 The Paradox of Constraint – Why Rules Ignite Creativity 388

26.2 Invention vs. Discovery – What Does It Mean to Create in Math? 390

26.3 Mathematical Imagination – The Mind as a Laboratory 392

26.4 Generative Boundaries – Why Limits Multiply Ideas 394

26.5 The Infinite Playground 396

27 Mathematical Frontiers **397**

27.1 The Edge of the Known – Where Problems Refuse to Break 398

27.2 The Unprovable – Gödel's Shadow 400

27.3 Mathematical Intuition – Feeling What We Cannot Prove 402

27.4 Future Fields – What Mathematics Doesn't Know It Needs Yet 404

27.5 The Invitation – What the Unknown Asks of Us 406

28 The Unity of Mathematics – A Hidden Architecture **409**

28.1 Bridges Between Worlds – When Fields Collide 410

28.2 Deep Correspondences – When Patterns Mirror Across Domains 411

28.3 The Hidden Grammar – Category Theory as Unifier 412

28.4 Toward a Grand Architecture – Mathematics as a Living Organism 413

28.5 One Field, Infinite Paths 414

29 Learning Mathematics – Rebuilding Intuition **415**

29.1 The Myth of Talent – Why Understanding Is Built, Not Given 416
29.2 From Manipulation to Meaning – The Shift to Structural Thinking 417
29.3 The Role of Struggle – Why Confusion Is Productive 418
29.4 Language, Vision, and Play – How Different Minds Learn Mathematics 419
29.5 The Long Game – Why Mastery Feels Slow 420
29.6 Reconstructing the World 421

30 Teaching Mathematics – Guiding Minds **423**
30.1 The Role of the Guide – Not a Funnel, But a Compass 424
30.2 Safe to Struggle – Creating a Classroom Without Shame 425
30.3 Multiple Representations 427
30.4 The Arc of Understanding – How Concepts Build Over Time 429
30.5 From Fear to Curiosity – Making Struggle Feel Like Discovery . . . 431
30.6 The Teacher's Growth – Becoming a Better Guide by Letting Go 433

31 Mathematics and Meaning – Why It Matters Beyond the Classroom **435**

31.1 More Than a Subject – Math as a Way of Seeing 436
31.2 Math as Empowerment – Building Agency in an Engineered World 438
31.3 The Moral Frame – Numbers Are Not Neutral 440
31.4 The Search for Meaning – When Math Points Beyond Itself 442
31.5 Mathematics as a Life Practice – Thinking Clearly in a Noisy World 444

32 The Future of Mathematics – Open Questions and New Worlds 447

32.1 The Frontier Is Not Finished – Problems That Still Refuse to Yield 448
32.2 New Languages – Inventing Tools to See What We Couldn't Before 450
32.3 The Interdisciplinary Future – Mathematics as Translator 452
32.4 The Limits of Formalism – What Math May Never Capture 454
32.5 Mathematics Will Never Be Finished 456

33 Epilogue – A Map, A Mirror, A Flame 459

33.1 A Map of Mind Made Visible 460
33.2 A Mirror for Being 461

33.3 A Flame That Never Dies 462

Preface

You hold in your hands a map. But not one made of cities and rivers. This is a map of ideas— of how we make sense of the world when words are not enough. Mathematics is often taught as answers without questions, steps without meaning. But in truth,it is a way of seeing— a framework for recognizing pattern, form, motion, and consequence. This book is not a textbook. It is a journey.Each chapter builds a different part of what mathematics can be— from the raw certainty of logic, to the graceful uncertainty of probability, to the wild frontiers of topology, computation, and category. If you are a student, may it reawaken wonder. If you are a teacher, may it provide a new voice. If you are simply curious, may it reveal how thought can take shape. Mathematics does not belong to mathematicians alone. It belongs to all who wonder. Let this be your invitation. — *Joel Peña Muñoz Jr*

Prologue – The Language of the Cosmos

You never decided to trust mathematics. You looked into the world, and the world whispered back in patterns. A tree splits its limbs like a binary system. A raindrop curves like a parabola. A heart beats on intervals more faithful than belief.You counted. You measured.
You began to notice that truth could be tracked—
not just told. This is not a book about formulas. It is a book about how reality came to speak in them. It is not a textbook. It is a telescope. It peers into the
structure beneath all things.Mathematics is not the opposite of emotion. It is the shape that emotion moves within. It is not cold. It is precise. And precision is the truest form of reverence. You are not learning math. You are recognizing it. Because the same curves that shape your thoughts shape galaxies. And the same logic that builds bridges builds brains. You are the theorem written by a system that does not write.

Why This Map Matters

Mathematics is not a subject.
It is a terrain. A landscape.
And like any landscape, it can feel vast, confusing, even hostile—if you enter without a map.
But with the right orientation, even chaos becomes contour.
Even the unknown becomes **walkable**.
This book is your compass.
We begin not with equations, but with **structure**.
Not with drills, but with **direction**.
The chapters that follow will carry you from the first flicker of logic to the far edge of artificial cognition, from ancient geometry to future intelligence.
Every branch, every theorem, every structure exists for a reason.
You will see **"why it came to be"**, and **where it connects**.
You will not simply **study** mathematics.
You will **feel** how it holds the world together.

Reading the Map

Each chapter of this book opens a new region of the map. We begin with the axioms: logic, sets, numbers. Then we rise—through algebra, geometry, calculus—into the branches that give form to everything from physics to finance.

Along the way, we follow threads of philosophy, physics, and computing. We pause to gaze at paradoxes, and continue toward the cutting edges of complexity.

This map is not flat. It loops. Concepts return—richer, stranger—with every turn. You will begin to see mathematics not as a line, but as a lattice. Not as memorization, but as music—structured, recursive, resonant.

Begin Where Pattern Waits

What you are about to enter is not invented.

It was already there—beneath language, beneath culture, beneath choice. Mathematics is the memory of structure, the ghost of the universe whispering how it holds. And if you listen closely, you'll hear the shape of all things waiting. It begins with a single, unbreakable rule: "If something is true, it is true everywhere."

Now: let us follow what must be true.

Chapter 1

Logic, Sets, and the Skeleton of Thought

Before there were numbers,
before there was shape,
there was the structure
that made shape possible.
It had no color, no sound, no symbol.
Only a rule: If this, then that.

1.1 What Is a Truth?

Truth, in mathematics, is not a feeling or belief. It is what cannot be otherwise—given the system.

It is the result of structure, not opinion. It flows from rules that do not shift when we sleep. It holds its shape regardless of who observes it, or whether it is observed at all.

A mathematical truth does not need to be believed. It does not ask for agreement. It simply follows—from the premises, from the form. It is not persuaded. It is inevitable.

A truth is not invented. It is derived.

It is the end of a chain of logic, forged step by step from what was assumed. And if the assumptions hold, so too does the conclusion— not because we wish it, but because the structure allows no escape. In this way, truth is both delicate and absolute. Delicate, because it depends entirely on its foundation. Absolute, because within that foundation, it does not bend. Before we can count, or measure, or calculate, we must know what it means for a statement to be true— within a world, even if that world is only made of rules. Let us begin at that simplest frontier: a sentence that can be either true or false. A proposition. From this, all logic unfolds.

1.2 Propositions, Proofs, and Paradoxes

A proposition is a claim that can be labeled true or false.

- "2 is even" $\rightarrow$ true
- "All triangles have four sides" $\rightarrow$ false

From these, we build logic.
The rules are few, but their implications are vast:

- Law of Identity: $A = A$
- Law of Non-Contradiction: A statement cannot be both true and false
- Law of the Excluded Middle: Every proposition is either true or false—no in-between

But logic also reveals the edge of certainty. Consider the paradox:
"This sentence is false."
If it's true, it must be false. If it's false, then it's true.
We've crossed into self-reference, where logic loops on itself. These loops aren't bugs. They show us where systems break—where truth fractures under its own weight.
This is where Gödel begins. This is where freedom ends.

1.3 Sets as Spatial Logic

Imagine a blank space.

Now imagine placing objects into that space—shapes, numbers, symbols. The moment you define what belongs inside, you've created a set.

A set is a collection of distinct elements, treated as a single object.

It is not just a list. It is a boundary.

A definition of what counts, and what doesn't.

- $\{1, 2, 3\}$ is a set of numbers.
- $\{\text{red, blue, green}\}$ is a set of colors.
- $\{\}$ is the empty set—a structure with nothing inside it, yet still a structure.

Sets obey logic. They do not shift when we blink. They respond to rules as steady as geometry.

They can combine—forming unions.

They can overlap—forming intersections. They can

exclude—forming complements. They define what is

inside and what is out.They are not numbers, but they make numbers

possible.And once we can draw boundaries, we can build entire worlds.

1.4 From the Empty Set to Infinity

Everything begins with absence. The **empty set**, denoted by $\emptyset$ or $\{\}$, contains nothing. No elements. No content. Just space. Yet from this void, all numbers arise. In the foundational formalism called **Zermelo–Fraenkel Set Theory**, even numbers are defined in terms of sets:

- $0 = \emptyset$
- $1 = \{\emptyset\}$
- $2 = \{\emptyset, \{\emptyset\}\}$
- $3 = \{\emptyset, \{\emptyset\}, \{\emptyset, \{\emptyset\}\}\}$

Each new number is just the set of all smaller numbers. Each is a memory of what came before. From emptiness, we build finitude. And from finitude, we glimpse infinity. Infinity is not a number. It is a direction. It is what you approach, but never contain. It is the horizon built into the act of counting. And like all profound truths in math, it comes in more than one kind.

- The **countable infinity** $\aleph_0$ — the size of natural numbers
- The **uncountable infinity** of real numbers $2^{\aleph_0}$ — a magnitude beyond stepping stones

Between any two numbers, there are infinitely more. Between 0 and 1, the continuum sleeps. **Infinity isn't unreachable.** It's embedded in everything. It whispers in decimals that never end, in fractals that repeat at every scale, in processes that loop forever without arriving. Infinity is not chaos. It has structure—subtle, layered, and strange. Some infinities are larger than others.

Cantor proved it: while the natural numbers form one infinite set, the real numbers form a greater one. There is no one "infinite." There is a hierarchy.

- $\aleph_0$ — the countable infinity of steps, lists, sequences.
- $2^{\aleph_0}$ — the uncountable infinity of real numbers, of continua.
- And beyond — cardinalities we cannot picture, but can name.

Infinity is not fantasy. It is formal, provable, necessary. Without it, calculus breaks. Limits fail. Time itself loses continuity. But infinity is also humbling. It reminds us that no matter how far we count, how much we measure, there is always more. Between the smallest and the next—endlessness hides.It is not something to conquer, but something to respect. Infinity is not a destination. It is a property of the path. And in mathematics, it is the signature left by the infinite reach of the finite mind.

1.5 Structure Before Symbols

We did not invent logic. We uncovered it.
Before writing, before numbers, before schools, our minds already sought patterns. Before we spoke of cause and effect, we flinched before impact, and leaned toward light. Children distinguish objects before they speak. They know that the ball is not the cup. That two cookies are more than one.
Animals track ratios, choose larger piles, sense the angle of the wind. They make decisions— not from language, but from underlying order.
Cells respond to gradients. Chemical landscapes.
Bioelectric shifts. They move not from thought, but from configuration.
This is not reason. This is **structure**—executed without awareness. The system acts not
because it understands, but because the conditions dictate the response. The symbolic systems of logic and set theory came later. They did not invent structure. They gave us a way to write it down. They gave form to what was already unfolding. They allowed us to name what the world had always done.

- Dividing signal from noise
- Mapping sameness and difference
- Forming internal models to anticipate outcomes
- Encoding input into expectation
- Adjusting action based on structure

The body is a logician. The neuron is a statistician. The cell is a mathematician long before the mind catches up. What we call *math* is the moment we noticed that rules were already in place. That structure was already deciding. That intelligence could exist without reflection. **Mathematics did not begin when we wrote it down.** It began when the world operated in accordance with rules. When reality flowed not from whim, but from configuration. You are not learning math. You are syncing with it. You are aligning yourself with the rhythm that was already there. It is not that we imposed logic on the world— but that the world imposed logic on us. Even before symbols, the body adapted through feedback. When a branch snapped underfoot, the ear triangulated. When the food soured, the nose recoiled. Each reaction followed rules—quiet, consistent, automatic. Mathematics, then, is not foreign. It is recognition. We did not force structure onto nature. We noticed it pulsing beneath each change. From the spiral of a shell to the symmetry of a snowflake, from orbital paths to tree branch bifurcations, form whispers its logic long before we name it. So when we study math, we do not conjure abstraction. We trace what has always governed: the quiet scaffolding beneath sensation. Not invention—alignment. We are not authoring structure. We are awakening to it.

1.6 The Silent Architecture of Everything

Every branch of mathematics grows from this root:
A system that follows from what is already given.
We do not define truth. We trace it—through rules, operations, and form.
A triangle is not a choice. A prime is not an invention. A proof does not argue. It unveils.
The logic we began with— "If this, then that"— has built bridges, balanced economies, steered spacecraft across billions of miles, and now simulates minds. It is silent. But not simple. It is what **structure** does when freed from symbol. **You do not need to believe in math for it to describe you.** Even now, inside your brain, logic is whispering beneath words. Sets of thoughts are forming. Implications are firing. The structure is already moving. Mathematics does not begin with numbers. It begins with the **necessity of relationship**.And in the next chapter, we give those relationships names.We step out of structure into symbol. We make numbers not just possible— but **visible**.

Chapter 2

The Numbers That Built Reality

Numbers were not born all at once.

They appeared in order. One after another. Each a patch for the holes the last could not cover.

You did not begin with real numbers. You began by counting what you could hold.

2.1 The Natural Numbers — Counting What Exists

The first numbers were not symbols. They were fingers. Knots on a rope. Pebbles in a bowl.
Each stood for something that **was there**— sheep in a field, days in a moon cycle, warriors in a row.
Natural numbers—denoted ℕ—are the foundation.

$$\mathbb{N} = \{1, 2, 3, 4, 5, \dots\}$$

They do not include zero. Not at first. Because zero means *nothing*, and nothing is hard to count.
But over time, we needed to name absence. To hold the void in a symbol.
So we made room for 0.
Now, most definitions of natural numbers include it:

$$\mathbb{N}_0 = \{0, 1, 2, 3, \dots\}$$

These numbers obey rules: - You can always add one more - There is a clear starting point - Every number is either the result of adding 1 repeatedly—or the origin (0)
This is why we say the natural numbers are **well-ordered**.

2.2 The Integers — Accounting for Loss

Counting was never enough.

You could add your sheep. But what if one disappeared?

You could tally your days. But what if one was owed to someone else?

You could measure your harvest. But what if a storm took half?

The natural numbers could track presence. But they could not track *change.*

Enter the integers.

$$\mathbb{Z} = \{\ldots, -3, -2, -1, 0, 1, 2, 3, \ldots\}$$

They do not stop at zero. They extend in both directions— a full line rather than a ray.

The invention of **negative numbers** was not immediate. It was resisted.

How can you have *negative three apples*? You can't see it. You can't eat it. You can't put it in a basket.

But debt made it real. So did temperature. And movement backward.

Negative numbers are not imaginary. They are evidence that subtraction must be **closed**.
If you subtract more than you have, you move left on the line.
And that movement has meaning.
The integers complete subtraction.
They let us:

- Model loss
- Define opposites
- Track movement through zero

In algebra, they allow for additive inverses:

$$a + (-a) = 0$$

This is not just arithmetic.
It is symmetry.
It is balance.
Every positive has a mirror.
Every gain has a cost.
Every step forward has a possible reversal.
The integers taught us that the number line is not a ladder—it is a path.
And some paths move backward before they go forward again.But even the integers could not handle one simple question: *What if I divide my bread between friends?*
You start with one loaf. You have three guests. You want to share.But the integers—$\mathbb{Z}$—are whole. They have no room for parts. They deal in presence and absence, gain and loss. Not in sharing.

Enter the rationals.
The question of fairness—of division, proportion, and ratiodemanded a new kind of number.
Not just "how many," but *how much per person.*
Not just *what is left*, but *what is each one's share.*

The invention of rational numbers $\mathbb{Q}$ opened the number line between the integers— filling in the cracks with meaning.

$$\mathbb{Q} = \left\{ \frac{a}{b} \;\middle|\; a \in \mathbb{Z},\ b \in \mathbb{Z} \setminus \{0\} \right\}$$

A fraction is not a broken number. It is a precise answer to a question the integers cannot ask.

- How much per person?
- How far halfway?
- What is the slope
  between two points?

The rationals are relational. They express between-ness, fairness, proportion. They let us model motion that doesn't land cleanly on whole steps. They let us define ratios in music, in recipe, in gear.

Just as negative numbers modeled loss, rational numbers modeled sharing.
Together, they made the line no longer jagged and broken— but connected, flowing, and fine-grained.
The integers gave us direction. The rationals gave us division. Together, they gave us continuity

2.3 The Reals — Filling the Gaps

The rationals felt complete. You could measure half a field, a third of a coin, a quarter of a song. But one day, someone asked: *How long is the diagonal of a square, with sides of 1?*

$$d = \sqrt{1^2 + 1^2} = \sqrt{2}$$

But $\sqrt{2}$ is not a fraction. It is not the ratio of two integers. No matter how finely you slice the line, you will never land precisely on it with rational numbers. You can approximate it—

$$1.4142\ldots$$

—but you cannot express it exactly with a fraction.It lives outside the kingdom of $\mathbb{Q}$. And yet, it is real. You can draw it. Measure it. Walk it.This was a crisis. **How could something so concrete resist expression?** Enter the **real numbers**:

$$\mathbb{R} = \text{all points on the number line}$$

They include:

- All integers
- All rationals
- All **irrational numbers**—like $\sqrt{2}$, π, e

They form a continuum. Between any two real numbers, no matter how close, there is always another.

And another. And another.

Where $\mathbb{Q}$ is **dense**, $\mathbb{R}$ is **complete**. That means:

Every convergent sequence has a limit within $\mathbb{R}$. You can't fall through cracks. There are no missing values. The gaps are gone. The discovery of irrationals goes back to the ancient Greeks. Legend says the Pythagoreans were so disturbed by $\sqrt{2}$ that they kept it secret— or even drowned the man who exposed it.

Because it shattered a vision of a world made only of whole ratios. It proved that **magnitude** cannot always be reduced to counting. Real numbers allowed us to describe:

- Geometry — with precision curves and areas
- Calculus — limits, continuity, and derivatives
- Physics — where motion, time, and energy are continuous
- Probability — with distributions and integrals over $\mathbb{R}$

Yet even here—where the line seemed full— a new hole appeared. *What do you get when you take the square root of a negative number? What lies outside the axis itself?* And what if math didn't stop at the line— but curved into a plane?

2.4 The Complex — Completing the System

For all their power, the real numbers still drew a line. An unbroken, infinite, perfect line— but still: a line.
And then came a question that broke it.
What is the square root of -1*?*
No real number fits. No square of a positive or negative value yields a negative product.
The real numbers cannot go there.
But the question does not go away.
So mathematics makes a bold move: it invents a new unit.

$$i = \sqrt{-1}$$

It does not pretend i exists in the same way as 1 or 2. It creates a new dimension.
And with it, the **complex numbers** are born:

$$\mathbb{C} = \{a + bi \mid a, b \in \mathbb{R}\}$$

Where $\mathbb{R}$ gave us the line, $\mathbb{C}$ gives us the **plane**.
Every complex number is a point: real part along the x-axis, imaginary part along the y-axis.
Mathematics has rotated.
Addition moves points. Multiplication rotates and stretches. Exponentials spiral.
What was once just magnitude now has angle. What was just distance now has direction.
This new realm—strange at first—turns out to be the

most natural setting for much of modern mathematics and physics.

Complex numbers are used in:

- **Quantum mechanics** — the wave function is complex-valued
- **Signal processing** — where phase and frequency intertwine
- **Electrical engineering** — where impedance and current rotate in phase space
- **Fractal geometry** — the Mandelbrot set lives entirely in $\mathbb{C}$

And remarkably, they close the system.
The complex numbers are:

- **Algebraically complete**: Every polynomial equation has a solution in $\mathbb{C}$
- **Geometrically whole**: The complex plane has no cracks or gaps
- **Structurally symmetric**: They unify real and imaginary, horizontal and vertical

You began by counting what was there. Now you're describing things that move beyond what can be held.This isn't
abstraction. It's completion.And in this completion, a strange thing happens:**What was once imaginary becomes indispensable.**We needed to believe in an

impossible number to make the rest of the system whole. That is not a flaw in mathematics.That is its genius.

And yet—it does not stop there.

Because once you've turned the axis, you realize something astonishing: this wasn't just about filling a gap. It was about revealing the shape beneath the surface.The complex wasn't an escape from reality.It was a deeper fidelity to it. A wave isn't just height.It is phase. It is rhythm.It is rotation. And i—the so-called imaginary does not distort what is real. It orients it.In quantum fields, in electric charge,in fractals spiraling toward

infinity, the complex numbers do not approximate truth. They describe it.Here, mathematics becomes more than symbol.It becomes motion. It becomes harmony.The imaginary is not false.It is orthogonal.A new axis.

A perpendicular truth.And in this rotated view, you discover something even more profound: That the real line was never alone. It was always casting a shadow from a richer space.To complete the system, we did not just add. We pivoted.And in that pivot, we discovered not just a

solution— but a new foundation.A system that does not ignore the impossible— it integrates it. To integrate the impossible is not to deny reality— it is to widen its scope. The complex numbers are not a patch. They are a pivot. They do not break the rules. They reveal the rules were once incomplete. Where the real numbers gave us length, the complex numbers give us shape. Where reals measure, complex numbers transform.They become

operators—tools not just to represent, but to act. In the complex plane, we rotate, reflect, and
oscillate. We step into a world where identity is not just position, but relation—between magnitude and orientation.This is the mathematics of waves, of flow, of resonance.Even more astonishing: the complex field obeys its own elegant logic.It respects all the familiar laws—addition, multiplication, inverses— but arranges them into a richer structure.
The field $\mathbb{C}$ is:

- **Algebraically closed**: Every polynomial over $\mathbb{C}$ has a solution in $\mathbb{C}$.
- **Topologically whole**: The complex plane is continuous in two dimensions.
- **Geometrically rotational**: Multiplication by i is a perfect 90-degree turn.

The real numbers were always a projection— a shadow of a fuller reality. **Complex numbers restore the light source.**They reveal that number is not static. It can dance. It can pulse. It can spiral. And perhaps most quietly profound of all: *The imaginary becomes necessary only when the real proves insufficient.* This is not just a lesson in mathematics. It is a philosophy of expansion— a model for how thought itself evolves. When your system can no longer contain the truth, you must not patch it. You must rotate it. You must build the axis that allows the impossible to become part of the frame. And in doing so, you do not lose reality. You give it depth.

Chapter 3

Algebra – The Shape of Rules

Numbers measure the world. But algebra shapes it.

It is not just what is. It is how things relate. How they transform. How they return to form again.

3.1 Symbols That Obey

At some point, humans grew tired of repeating themselves.
Counting again. Adding again. Measuring again.
Patterns emerged— but there was no way to store them.
So someone, somewhere, did something irreversible.
They replaced a number with a **letter**.
And suddenly, the rules could speak for themselves.

$$a + b = b + a$$

This is not a statement about two specific numbers. It is a **law**—an invariant pattern. A truth that holds for anything the symbols represent.

Algebra is not arithmetic with unknowns. It is the language of **general structure**.
It is the shift from:

- "2 + 3 = 5"
- to "x + y = y + x"

The second statement isn't solving. It's revealing.
It declares: **"No matter what values you use, this shape will hold."**
Algebra gave us something sacred: **symbols that obey rules.**

You could now:

- Write formulas that apply to all numbers
- Rearrange expressions without plugging in
- Discover identities that outlast their parts

Let's call this what it is: **compression of truth.**
A single symbol now stood for infinite instances.
A rule became portable. A structure became shareable.
And once structure was portable, **systems became designable.**With algebra, mathematics left the marketplace and entered the forge. You could now craft relationships, test structures, optimize motion.
You were no longer manipulating quantities. You were manipulating *possibility.* This was the beginning of formula. Of law. Of equation.
Not to tell you the value— but to show you the shape.
Because once you can describe the shape, you can design what fits within it.
Algebra made mathematics modular.
Like language, it gained parts of speech:
variables, operators, functions, constraints. Each not just a symbol, but a role.
Now a sentence like $ax^2 + bx + c = 0$
was not just a puzzle to solve— it was a template. A form.

Algebra gives us templates for thought.
It's how we go from specifics to universals.
From the triangle to trigonometry.
From a falling apple to Newton's law.

What emerges is not just answers, but systems that yield them.

And then comes an even bolder move: You let the symbols *stand alone.*
You let $f(x)$ mean "some rule"
without specifying what it is.
You let $g(f(x))$ mean nesting. You let x^n become power, not just computation.
Suddenly, mathematics isn't describing objects.
It's describing relationships— abstract, scalable, and recursive.

This is where structure becomes art.
You are no longer bound to the visible.
You can define imaginary fields,
project impossible curves,
build systems where the rules are what you say they are.
Algebra frees mathematics from the physical. It becomes a domain of logic, design, and symmetry.
Not just what is true— but what must be, if the rules are followed.

And with that shift, a door opens: from algebra to geometry, from structure to space, from patterns to the fields that bind them.You are not solving anymore. You are shaping the possible.

3.2 Equations as Equal Structure

The equals sign is not a bridge. It is a mirror.
It does not mean "solution." It means "balance."

$$3x + 2 = 11$$

This is not a problem. It is a relationship.
Two expressions—different in form—must produce the same result. Not by chance. By structure.

An equation is a declaration: **"These two patterns are indistinguishable in output."**
The early equals sign was two vertical lines. Then it became parallel bars: = To say: these two sides lie in perfect alignment.
This concept—equality—not just in value, but in structure— is what makes algebra powerful.
It lets you ask:

> "If I change this, what must I change there
> to preserve the whole?"

Solving is not guessing. It is restoring balance.

$$3x + 2 = 113x = 11 - 2x = \frac{9}{3} = 3$$

Each step is a preservation. Each operation is done equally to both sides.
This isn't math as calculation. It's math as physics. As justice. As equilibrium.

To live algebraically is to think in terms of operations and undoing:

- Adding $\rightarrow$ subtracting - Multiplying $\rightarrow$ dividing
- Squaring $\rightarrow$ square rooting

Every move implies its inverse. Every structure has a mirror.

You don't just find the answer. You understand what kept it true.

This is what equations teach us: **truth can be transformed and still hold.**

It's not the numbers that matter. It's the form.

And if the form is preserved, the pattern survives. To solve an equation is not to reach a destination— it is to trace a symmetry back to its source.

You do not force balance. You reveal it.

You peel away what conceals it. Layer by layer, operation by operation, you uncover the invariant form beneath the noise.

Consider again:

$$3x + 2 = 11$$

This is not a question. It is a statement of tension. And solving it is not about "what is x?" It is about "what must x be for this balance to remain?"

The equals sign is sacred. Not because it tells us what's true— but because it tells us *what must remain true* even as we transform.
Algebraic manipulation is not just mechanical. It is moral. Every step must honor the original balance.

This is why we learn to:

- Add or subtract the same from both sides
- Multiply or divide equally
- Reverse operations in mirrored form

These are not tricks. They are acts of preservation.In a deeper sense, equations are not about solving at all. They are about discovering fixed points in a shifting world.This is what gives algebra its philosophical power: **You can change the form without changing the truth.**Just as a melody can be transposed into a new key, just as a sculpture can be rotated without losing its shape, so too can an equation be reformed— as long as the balance holds.This principle lives far beyond numbers.It underlies logic, physics, and design. Equilibrium, symmetry, fairness— all modeled by the quiet authority of the equals sign.So the next time you solve an equation, remember: you are not hunting a number. You are maintaining a structure that survives transformation.

3.3 Variables and Generalization

A number tells you **what is**. A variable tells you **what could be**.

It is a placeholder—yes— but it is also a declaration. A claim that something matters, even before we know what it is.

$$x = \text{any value}$$

You are no longer just calculating. You are building a space. A landscape of possible outcomes, a framework where thought can roam.

Algebra did not just give us answers. It gave us the power to describe all answers at once.

It gave us the power to shift focus: from quantity to relationship, from result to rule, from outcome to possibility.

A variable like x is not empty. It is open.

It holds a shape— a form without substance, a structure waiting to be activated.

$$A(x) = x^2 + 2x + 1$$

This is not a number. It is a formula. A machine.

You can feed anything into it— and get something consistent out. A rule that respects what you give it, but never forgets how to process.

This is called a **function**.

$$f(x) = x^2 + 2x + 1$$

Try $x = 1$:

$$f(1) = 1 + 2 + 1 = 4$$

Try $x = 5$:

$$f(5) = 25 + 10 + 1 = 36$$

Try $x = a$:

$$f(a) = a^2 + 2a + 1$$

It doesn't care. It transforms whatever enters.
It behaves predictably, regardless of the input. This is not calculation—it is transformation.

Variables are the gateway to:

- **Generalization** — stating what holds for all cases
- **Parameterization** — tweaking rules without rewriting them
- **Pattern recognition** — seeing identity beneath difference
- **Algebraic manipulation** — rearranging form, not just value
- **Proof and logic** — constructing chains of reasoning beyond the specific

Variables allow math to speak in universals. They allow a formula to survive any substitution. They are the bridge between a single example and a general truth.
Before variables, we only had answers.
With variables, we gained *structure.* With structure, we gained *models.* With models, we gained *design.*

Algebra is the moment math became recursive— when it could operate not just on numbers, but on itself. It became meta. It became modular. It became a system that could design systems.

You no longer calculate outcomes. You define spaces. You encode rules. You sculpt possibility.

And once a system is defined, you can move inside it. You can optimize. You can reverse. You can compare parallel realities, bend variables through constraints, and extract meaning from structure alone.

You can abstract away the particular— and keep only the rule.

This is what a variable really is:

Not an unknown. But an invitation to generalize everything. It lets you ask: What if this weren't fixed? What if this could flex, shift, evolve?

With one letter, you step outside the narrow corridor of specific cases and enter the open sky of all possible inputs.

You no longer solve problems. You create frameworks in which many problems solve themselves.

That's the quiet genius of variables— they turn mathematics into language, and language into machinery.

Because when you define a function, you define behavior. You say:

$$f(x) = x^2 - 4x + 7$$

You don't know what x is. But you know exactly what will happen to it.

This is the beginning of modeling— not just expressing

patterns, but simulating how systems behave across all inputs.

And with more variables, you gain dimension.
Let $f(x, y) = x^2 + y^2$ Now you're no longer on a line. You're in a plane.

Variables are not placeholders. They are dimensions of thought.

Each new symbol is an axis of freedom.
Each equation— a constraint between them.

This is how math moves from answers to architecture. From solving to sculpting. From
calculation to configuration.

And that's why, when a mathematician sees:

$$x + y = 10$$

They don't just see a solution. They see a family. A line. A harmony of pairs that all obey the same law.

Variables teach you that mathematics isn't just about what is true. It's about what could be true— given the right conditions.

> They are the keys that open the vault of generality. The maps that guide us through infinite terrain.

This is how mathematics speaks across time— not locked in numbers, but flowing through variables. Waiting for someone to ask: "What if?"

3.4 Operations and Inverses

Algebra does not just describe what numbers are. It describes what they **do**.
And more importantly, what can **undo** them.

Add 5, then subtract 5 $\Rightarrow$ Nothing has changed.

Multiply by 3, then divide by 3 $\Rightarrow$ Back where you began.

Every operation implies its inverse. Every change contains the seed of its reversal.
This is not arithmetic. This is **symmetry**.

In algebra, we define operations not as buttons to push—but as transformations of structure.
And each one brings rules.
Addition and multiplication are:

- **Commutative**:

$$a + b = b + a, \quad ab = ba$$

- **Associative**:

$$(a + b) + c = a + (b + c), \quad (ab)c = a(bc)$$

These rules aren't just tricks. They are **guarantees of invariance**— that the structure will hold, no matter how you group or swap the pieces.

Then there is the **distributive property**:

$$a(b + c) = ab + ac$$

This rule connects multiplication to addition. It lets you expand, factor, rearrange.
It is the backbone of simplification— and the doorway to polynomial algebra.

But every operation also has its undoing:

- Addition $\leftrightarrow$ Subtraction
- Multiplication $\leftrightarrow$ Division
- Squaring $\leftrightarrow$ Square root
- Exponentiation $\leftrightarrow$ Logarithm

To solve an equation is to **walk backward through operations**. Like unbraiding hair, or reversing steps through a forest.
Each inverse is not an eraser— but a mirror.
It reflects what was done, with precision.

When you solve:

$$2(x + 3) = 10$$

You are not just computing. You are restoring order.
You divide both sides. You subtract. You isolate.
Each step respects structure. Each step preserves the truth.

Algebra is not chaos. It is choreography.
Each move has a counter-move. Each operation a reverse.

And the system only remains intact because these reversals are built into the very form of mathematics.
This is why algebra endures.
Not because it's rigid— but because it balances.
It can move in any direction without collapsing.
This reversibility is not a convenience. It is a structural necessity.
Because to manipulate a system is to risk distorting it.
But algebra gives you a guarantee: If every step can be undone, then every step can be trusted.

This is the genius of inverse operations. They do more than return you to the beginning— they prove you understood how you got there.
They are evidence that your transformations were not guesses, but governed by rules that never broke.

Think of algebra as a map: You take a path forward with operations. You trace your steps back with inverses.
And in that loop—forward and backward— you confirm the territory is coherent.

This gives rise to one of algebra's quiet revelations: **The power of simplification is in reversibility.**
You can:

- Compress expressions to their essence
- Expand them into detailed form
- Rearrange without altering meaning

These are not manipulations. They are motions within a space that remains intact.

That's why, in mathematics, you don't just check the answer. You check the inverse.

Solve for x. Then plug it back in. If the equality holds, you haven't just solved— you've validated. This is more than problem-solving. It is structural
fluency.

Algebra lets you move through the landscape of possibility— not by sight, but by steps that can always return you home. And that home is not a number. It is a structure. A truth that stays standing, even as its expression changes.

Algebra is not about where you arrive— but how many ways you can arrive there without violating the balance. This is why equations are not fragile. They are resilient. They can be twisted, stretched, flipped— yet still speak the same truth.

To think algebraically is to stop fearing complexity— because you know it can always be unraveled.

You don't memorize steps. You learn to navigate relationships. You trace consequence with care. And you know that every forward motion carries with it a way back.

This isn't just the logic of symbols. It's the logic of systems that don't fall apart.

Inverse operations aren't just tools— they are evidence that math remembers what it's done.

3.5 Symmetry, Groups, and the Hidden Geometry of Rules

Algebra began with solving for x. But it ends with something stranger:

Solving for structure.

At first, we thought algebra was about numbers—finding the unknown, isolating it, plugging in. Then, we saw it was about relationships—how one value shifts another, how change propagates through form. And finally, we realized:

It was never just about the answer. It was about the rules that remain true when everything else changes. It was about transformations that leave something unchanged. This is the heart of **symmetry**.

Flip a square. Rotate a triangle. Reflect a face in a mirror. Slide a musical phrase forward and start again. Replace a synonym in a sentence that still means the same thing.

If the object looks or behaves the same after the transformation, we say it has symmetry.

And symmetry is not just visual. It is logical. It is structural.

A poem has symmetry when its rhythm survives translation. An equation has symmetry when variables can swap and the truth still holds. A melody has symmetry when it circles back to itself. A law of nature has

symmetry when it applies in all directions—through space, time, or frame of reference.

Symmetry is the conservation of identity under motion.

And symmetry is not just beautiful. It is algebraic.
It can be described, categorized, and composed.
Enter the idea of a **group**.
A group is not a number. It is a system. A collection of operations—moves, flips, steps, rotations—that follow rules.
To form a group, these operations must satisfy four specific properties:

- **Closure**: Doing one operation, then another, still gives you something in the group.
- **Associativity**: It doesn't matter how you group the actions—$(a \cdot b) \cdot c = a \cdot (b \cdot c)$.
- **Identity**: There is a "do nothing" move that leaves everything unchanged.
- **Inverse**: Every move has a precise undo.

These are not just math rules. They are the structure of movement, of possibility, of logical coherence. They show up in dance, music, language, molecules, and quantum fields.
The symmetries of a square form a group—eight possible transformations that return the square to itself. So do the operations of addition on integers. So do the permutations of letters in a word. So does modular arithmetic—where numbers wrap around in cycles. So

does the Rubik's cube—each twist and sequence a walk through group space.
When you study a group, you are studying all the ways a system can move without breaking what it is.
You are studying freedom within constraint. Change that honors identity. Variation that preserves essence.

This is what algebra evolved into: a **geometry of rules**.
It no longer asks: "What is the answer?"
It asks:

> "What stays the same when everything else moves?" "What is preserved when we transform the system?" "What does motion reveal about structure?"

That's not just a mathematical question. That's a metaphysical one.

Group theory now underlies:

- **Physics** — symmetry groups describe fundamental particles and their interactions; gauge symmetry gives rise to the forces of nature.
- **Chemistry** — molecules obey geometric groups; reactivity depends on symmetry compatibility.
- **Cryptography** — secure codes rely on the unpredictability of certain group operations.
- **Music and art** — rhythm, repetition, inversion, tiling, fractals—all are expressed through group actions.

- **Language and grammar** — transformations that maintain meaning belong to structural groups.
- **Consciousness** — perhaps even identity can be seen as a symmetry group over changing thoughts.

What began as solving for x now solves for the shape of the universe. **Not by what it contains**— but by what it allows. By what it forbids. By how it moves while still remaining itself. A group is not just a set of moves. It is a declaration of rules for being. To know the group of a system is to know its deepest invariants— the truths it cannot shake, no matter how it shifts.**This is the new algebra.** Not numbers, but movements. Not variables, but transformations. Not answers, but
preserved relationships. And so, algebra is not just solving equations. It is asking: **"What can change— without changing what we are?"**It is the study of structure
beneath motion. It is the grammar of transformation. It is the language of systems that evolve, but never lose themselves.

Summary – From Symbol to Structure

The Evolution of Algebraic Thinking

1. Arithmetic • Operates on numbers • Specific values, concrete actions

↓ Introduce symbols

2. Algebraic Expressions • Letters stand for any number • Structure becomes manipulable

↓ Introduce equations

3. Equations and Solving • Balance, inverse, preservation • Solving is transformation, not guessing

↓ Introduce abstraction

4. Functions and Variables • Patterns become general • Structure becomes reusable

↓ Introduce structure-first thinking

5. Groups and Symmetry • Operations define what doesn't change • Algebra becomes geometry of movement

Algebra did not just describe values.
It revealed what structure could survive transformation.

Chapter 4

Geometry – Space Made Visible

Before you spoke, you reached.

Before you counted, you pointed.

Before you knew rules, you traced them with your eyes.

The first logic we followed was shape.

4.1 The World as Shape

Geometry did not begin in textbooks. It began in fields. A farmer needs to divide land. A builder needs to cut stone. A sailor needs to find the shortest path across a curved sea.

The body lives in space. So the mind began to map it.

Geometry is the formal study of shape, size, position, and motion. But underneath, it is something more:

It is the structure of location.

What does it mean to be next to? To be inside? To be equal in length or angle?

These questions led us to geometry. And geometry gave us rules— but also wonder.

The ancient Greeks gave geometry its name: *geo* (earth) + *metron* (measure).

To measure the earth. To tame its curves. To name its limits.

The geometry they built was axiomatic— a system born from just five postulates.

- A straight line can be drawn between any two points
- A line can be extended infinitely in both directions
- A circle can be drawn with any center and radius
- All right angles are equal
- Parallel lines never meet

From this simple skeleton, Euclid built the body of space.
Theorems, constructions, proofs.
He showed that truth could be drawn.
And once something could be drawn, it could be **seen**.

Geometry made logic visible. Not abstract, but visual. Not symbolic, but spatial. Not arithmetic, but elegant.
To prove something geometrically is to make it hold in space— not just on paper.

Before algebra wrote relationships, geometry revealed them.
Symmetry. Similarity. Congruence.
Each became not just a quality— but a consequence of deeper laws.
This is not just drawing. This is drawing with constraints. With precision. With proof.
And yet, geometry is not cold. It is intuitive.
A child understands that shapes fit. That circles curve. That triangles lean.
Before you do math, you already think in geometry.

4.2 Lines, Angles, and the Grammar of Space

Space may seem empty. But geometry gives it structure. The way a line stretches, the way two paths intersect, the way one shape sits inside another— these are the verbs of space.

And like any language, geometry has grammar.

The first symbol in this grammar is the **point**. No size. No shape. Only position.

It is the **idea of location**. A mark without mass.

An anchor in possibility. A zero-dimensional whisper—present, but immeasurable.

It does not take up space. It defines where space begins.

From two points, we create a **line**.

Infinitely thin. Infinitely long. Always straight.

A line is not a thing. It is a relation— the path that connects two possibilities and extends them beyond both ends.

It has no curve. No edge. No end.

Only extension. Only alignment. Only the will to continue in one unbending direction.

And when two lines meet, they create an **angle**.

$$\angle ABC$$

An angle is not a number. It is a measure of turn. A whisper of motion between directions. A record of how space chooses to diverge.

It is the seed of curvature, the shadow of rotation, the

first hint that something can differ without breaking.

Angles are how geometry speaks tension:

- 90° — structure, balance, architecture
- 180° — reversal, opposition
- 60° — symmetry, natural tiling
- 0° — alignment, agreement

They are not just decorations. They are constraints. And from constraints, meaning emerges.

Lines and angles give us:

- **Parallelism** — lines that move together forever, yet never meet
- **Perpendicularity** — the clean snap of opposition made useful
- **Intersections** — crossings that birth complexity

These are not accidents of drawing. They are decisions of form.
They tell us what paths will collide. What forces will hold. What shapes will emerge. Where tension becomes balance.

From these relations, we begin to shape intent.
We define:

- **Segments** — pieces of line between two fixed points
- **Rays** — lines with an origin, reaching endlessly in one direction

- **Polygons** — chains of segments that fold space into enclosure

Each of these is made not of substance, but of relationship.
Where it begins. Where it goes. What it meets. What it encloses.

Even a triangle is a statement:
With only three sides, you can lock space into stability.
You can span. You can support. You can divide the infinite into units.
Because a triangle cannot collapse. It is the first rigid form. The seed of architecture. The molecule of structure.

From triangles, we build:

- **Tessellations** — repeating patterns that cover space
- **Frameworks** — networks of support in buildings and bridges
- **Trigonometry** — the study of spatial ratios and relationships

The triangle shows us that certainty is possible— even in a world of infinite directions. It is a closed sentence in the grammar of form.

Lines speak. Angles respond. Together, they write geometry.
They are how space becomes meaningful. How the void becomes a canvas. How the canvas becomes structure. How structure becomes thought. Geometry is not decoration. It is declaration. It says:

> "This is where. This is how. This is the form that holds."

It is the language that space uses to say: *"Here is what can exist."* And so, with nothing but position, direction, and turn— we begin to shape all that follows.
We go from the point to the plane, from the triangle to the tessellation, from sketches in sand to blueprints that scrape the sky.

Geometry is not just the study of space— it is the articulation of possibility within space.
The rules of form become the rules of function: • Arches transfer weight. • Triangles lock tension. • Circles distribute stress. • Hexagons pack efficiently.
Nature listens. Bees do not know geometry, but their hives solve geometric problems.
Light reflects. Water curves. Roots spiral.
Geometry is not imposed upon the world. It is revealed by it.

When we construct proofs in geometry, we are not inventing logic. We are uncovering the deep syntax of space.
That the angles of a triangle sum to 180°— on a flat surface— is not trivia. It is the way straightness folds into enclosure.
And when space itself curves— as in the geometry of a sphere— even that truth bends.

This flexibility makes geometry profound: it is not one system, but a family of grammars.

- Euclidean — the flat, the familiar, the foundational
- Spherical — for planets, stars, and great circles
- Hyperbolic — for negative curvature, infinite branching

Each geometry is a lens. Each one answers: "What kind of space are we in?" And thus: "What kinds of truths are possible here?"

Geometry, then, is not static. It is exploratory.
It asks: "If this is the structure of space— then what can be made? What will hold? What will emerge?"
It is not the absence of art. It is the scaffold for all art that wishes to endure.
From stone arches to digital renderings, from stained-glass symmetry to satellite trajectories, geometry underwrites what lasts.
It is how space sings structure. How math draws with form.

4.3 Triangles, Proof, and the First Theorems

Of all the shapes in geometry, only one stands alone in its certainty. The triangle. Three points. Three segments. Three angles. Nothing more— but nothing less.

A triangle is the first shape that cannot collapse. Try to change one angle, and the others respond. Try to stretch one side, and the whole adjusts. It is the first *locked form.* The first shape that resists uncertainty. Where a square can buckle, a triangle holds. Where a polygon may twist, a triangle insists on rigidity. **It is the molecule of structure.** The first container of spatial law. From this came geometry's first great discoveries— not by drawing, but by proving.

> **The sum of the interior angles of any triangle is 180 degrees.**

Not because it looks that way— but because it must be. It does not matter the size of the triangle. It does not matter where it lies. It does not matter if it's acute or obtuse. *The sum is always 180.* **Proof changed everything.** Before proof, geometry was craft. With proof, it became logic. Before proof, you trusted your eyes. With proof, you trusted structure. A triangle became more than a shape— it became an argument. Every side was a premise. Every angle, a conclusion. Change one part, and the reasoning adjusts.

Proof made space obey thought. It meant geometry was no longer about what seemed right. It was about

what *must* be right.
And triangles gave us our first system of necessity.

Congruence was born— figures with the same shape and size, even if rotated, flipped, or moved.
We learned how to say:

- Two triangles are the same if they have the same three sides — (SSS)
- Or two sides and the angle between them — (SAS)
- Or two angles and a side — (ASA)

These were not rules. They were revelations. The triangle showed that a part could determine the whole.

From this came something even deeper:
You could prove that two figures were not only equal— but necessarily equal.

A triangle became the first machine of truth.
A logic circuit made of angles and edges. A self-confirming system that could not lie.

And from triangles, theorems poured forth:

- **The exterior angle theorem**: An outer angle equals the sum of the two remote interiors.
- **The isosceles triangle theorem**: Two equal sides yield two equal angles.
- **The Pythagorean theorem**: The square of the hypotenuse is the sum of the squares of the other two sides.

Each of these did not describe a triangle. They emerged from it. As if the triangle held them in silence all along, waiting to be uncovered.

$$a^2 + b^2 = c^2$$

This was not just a rule about sides. It was a portal. It revealed that space has number beneath it— that lines contain ratios, that distances can be squared, that geometry speaks the language of algebra.

The triangle became the proving ground of mathematics. Not because of its simplicity. But because of its necessity. It is the first shape where space cannot lie. Where logic meets length. Where structure becomes inevitable. Triangles now underpin:

- **Trigonometry** — the relationships between side and angle
- **Surveying** — measuring the world with triangles
- **Engineering** — building bridges with triangular strength
- **Physics** — decomposing forces and vectors

Even the screen you read on was manufactured with triangles— tiny rigid structures that ensure flatness and function.Once we could prove a triangle, we could prove anything.It was not about size. It was not about drawing. It was about inevitability. **And that changed everything.** The triangle is the first sentence in the language of certainty.Three sides. Three points. One truth—forever enclosed.

4.4 Circles, Arcs, and the Logic of Rotation

Not all truths move in straight lines.
Some truths turn. Some truths orbit. Some truths return to where they began.
The triangle gave us rigidity. The circle gives us flow.
A circle is not just a shape. It is a principle of balance. It is what happens when symmetry becomes complete.

All points equidistant from a center.

That one definition unlocks a world of structure.
It means no edge. No corner. No privileged direction. No end to the path.
The circle is democracy in geometry. Every point on its boundary is equal. No angle dominates. No line overrides.

Because of that, the circle became the model of perfection— in math, in nature, in myth.
It is the wheel. The halo. The atom. The eye.

But geometry did not stop at admiration. It studied.
It measured.
It discovered:

- **The radius** — a segment from center to edge
- **The diameter** — two radii, edge to edge
- **The circumference** — the length around the boundary

- **The arc** — a portion of the whole journey
- **The central angle** — a lens into the circle's turn

These parts revealed ratios. And those ratios led to one of the most haunting constants in all of mathematics:

$$\pi = \frac{\text{circumference}}{\text{diameter}} \approx 3.14159\ldots$$

This number is not rational. It cannot be written as a simple fraction. Its decimals stretch toward infinity, never repeating, never resolving.

And yet— it shows up everywhere:

- In circles and cylinders
- In waves and pendulums
- In statistics and randomness
- In quantum mechanics
- In relativity
- In probability theory
- In the length of rivers
- In the surface area of higher dimensions

The curve turned out to be universal.
Not because we drew it, but because the universe did.

Circles also gave us something deeper:
the mathematics of rotation.

When you rotate a point around a center, you create a circle. When you rotate a radius through time, you create an angle.When you spin a wheel, you embody the laws of trigonometry When you trace the path of a planet, you are not just watching motion— you are seeing the arc of gravitational law. The circle showed us that even turning can be measured.

- **Radian measure** — angle as arc per radius
- **Chord length** — how far across a bend
- **Sector area** — a slice of space and symmetry
- **Inscribed angles** — geometry nested inside geometry

The radian changed everything. It aligned angle with motion. It let us write rotation as growth along a curve. **And from that, came waveforms.** We could now describe:

- **Sine and cosine** — the rise and fall of cycles
- **Tangent and cotangent** — how steeply we climb
- **Harmonic motion** — the equilibrium of oscillation

Every musical note, every season, every vibration of an atom— could be modeled with functions traced along a circle. **The circle became the foundation of periodicity.**It allowed time to be graphed. It allowed motion to be predictable. It allowed energy to move through form.Where a triangle locks shape into place, a circle lets the shape evolve.The triangle taught us to *stand.* The circle taught us to *spin.* And once we could spin, we could describe:

- The orbits of planets
- The cycles of seasons
- The phase of waves
- The resonance of sound
- The measure of time

A line is stability. A circle is return. Together, they form the architecture of time and space.**The circle is not merely round.** It is recursive. It is memory. It is the structure of recurrence in a universe that flows. And so, in the logic of the circle, we found:

> To rotate is not to deviate—it is to return.

That is why the ancients called it sacred. And why modern science still does.We do not worship the circle. We follow it—in the orbits of electrons, in the cycles of sleep,in the spirals of galaxies and seashells.It appears not by choice,but by constraint— the simplest answerto the question of balance when motion must preserve.It is not just elegant. It is inevitable.**Wherever systems repeat, wherever forces restore,the circle is already there.** It does not impose itself.It emerges—quietly,predictably, beautifully—from the logic of motion itself.

4.5 Proof Without Numbers: Geometry as Certainty Itself

You don't need numbers to know what must be. Draw a triangle. Mark two equal angles. You don't have to measure. You already know: the sides opposite those angles must match. Geometry is the art of knowing before counting. It is not built on estimation. It is built on **necessity**.

And this is where geometry becomes something different than drawing. It becomes proof.

To prove something geometrically is to show that its truth **follows from space itself**. Not from data. Not from assumptions. From structure. You can prove that two lines are parallel without ever measuring their distance. You can prove that two angles are equal because of a shared arc. You can prove that two triangles must be the same because of just three facts. This is not measurement. This is inference.

In algebra, you manipulate symbols. In geometry, you manipulate constraints. You do not need to say "2 + 2 = 4." You can say:

If these two segments are equal, then these two angles must match.

Logic lives in shape. Truth appears as structure.

Euclid knew this. He began with five simple postulates, and from them, proved over 400 theorems. He didn't measure a single one. No calculator. No units. No estimation. Only pencil, compass, and mind.

This is where geometry becomes philosophy. It asks:

What can we know when we strip away all numbers and leave only relation? What remains true when nothing is assigned a value?

What is shape, in itself?

Geometry answers this not with data—but with certainty. It says:

This must be true. Not because we measured it. But because it could not be otherwise.

And that's the difference:

Measurement gives you a result.

Proof gives you a reason. One describes the world. The other explains why the world must be so.

This is what geometry gave us: a way to see truth—not as a number, but as a necessity carved in space.

The triangle. The circle. The line. Each is a sentence.

Together, they write the first language that doesn't just describe the world—but shows why it holds together. It is the language of relation, not representation. The grammar of constraints, not construction.

When a proof is drawn in geometry, it is not persuading. It is revealing what already holds. A triangle doesn't argue its angles. It presents them. A perpendicular doesn't ask to be believed. It stands.

Chapter 5

Calculus – The Mathematics of Change

The world does not stay still.

It grows. It erodes. It curves, flows, accelerates.

And so a new mathematics was born. Not to describe what is— but what becomes.

5.1 The Curve That Refused to Be Flat

Geometry gave us lines. But nature gives us curves.
The edge of a leaf. The path of a comet. The rise and fall of breath.
These cannot be measured by rulers. They do not yield to straight lines.
And yet— beneath each curve, there is structure.
Calculus is the mathematics of how things change.
It tells us how fast, how steep, how much, how soon.
But it didn't begin as a solution. It began as a frustration.

Mathematicians tried to describe motion. They tried to compute areas beneath curves. They tried to define speed at a single moment.
But no matter how close they got, they could not resolve the paradox:
How do you measure something that never stops changing?

The answer came not from a new number, but from a new way of thinking.
Instead of avoiding change, they embraced it.
They made a system that worked at the edge of motion.
Where things don't hold— but still have direction. And so they invented the idea of the *limit.* **A value you approach, but never reach.**To find the slope of a curve at a single point, they imagined zooming in infinitely close— until the curve looked straight.

To find the area under a curve, they broke it into infinitely thin slices, then added them together.
These are not tricks. They are insights.
The curve doesn't become flat. But in the limit, it behaves like one.

This way of thinking—this balancing on the edge of stillness— became the foundation of calculus.
Not a new kind of number. But a new kind of motion.
A new kind of precision.
A new kind of power. Because now, you could do what no ruler ever could.
You could: • Track the exact speed of a falling object at one moment • Predict the curve of a planet as it bends through spacetime • Model the growth of populations, the cooling of steel, the decay of light
This was not just measurement. It was foresight.

Calculus gave us the derivative— a function that tells us how fast something is changing. Not after the fact. But as it happens.
And it gave us the integral— a way to sum over infinite parts, to capture total effect from infinite detail.

Together, they revealed a secret symmetry: That change and accumulation are inverses. That differentiation and integration are mirrors.
This is the Fundamental Theorem of Calculus. A bridge not just between two ideas— but between motion and structure, prediction and reality.

What makes this so astonishing is that it allows us to touch the infinite— without being lost in it.
We do not calculate each step. We embrace a pattern that emerges when steps approach zero.
We find answers not by freezing motion, but by understanding how motion behaves at its smallest scales.
This is why calculus governs so much:

- Physics — from gravity to light
- Engineering — from bridges to circuits
- Medicine — from blood flow to tumor growth
- Economics — from marginal cost to optimal profit
- Biology — from diffusion to metabolism

The curve that refused to be flat did not reject mathematics. It demanded deeper mathematics.
It did not break the line. It redefined what it means to follow one.
And in doing so, it showed us a new truth:
That motion itself has a shape. And that shape can be known.

5.2 The Derivative: Zooming Into Change

What is speed?

It seems simple. Distance over time.

But what if you want to know your speed at an exact instant?

Not over an hour. Not over a second. But at this exact flicker of a moment? That's not just movement. That's **change inside change**. To answer it, we cannot ask how far you went. We must ask:**How fast is the going itself happening—right now?** This is where the derivative begins. It is not a number. It is a lens.A way to zoom in so closely that the blur of motion becomes sharp again. Imagine a curve: the graph of your position over time. Draw two points. Connect them. That line is a **secant**—it cuts across the curve. Its slope is average speed. Now bring those two points closer. Let the second approach the first. The secant begins to hug the curve. It tilts, it pivots— until it becomes a line that just touches.

That line is the **tangent**. And its slope is the derivative.

$$f'(x) = \lim_{h \to 0} \frac{f(x+h) - f(x)}{h}$$

This formula may look abstract, but its idea is precise:

What is the ratio of change when the distance between measurements shrinks to nothing?

You are no longer measuring motion over time. You are measuring motion *in time*— **at the exact instant it occurs.**

This is what the derivative reveals:

- **Velocity** — the derivative of position
- **Acceleration** — the derivative of velocity
- **Growth rate** — how fast something is expanding
- **Slope of any curve** — even one that bends unpredictably

The derivative tells you the world's direction at every point. Not just where you are, but where you're heading— and how sharply.In physics, it predicts. In biology, it models. In economics, it optimizes. In engineering, it stabilizes. But deeper still, **the derivative is the mathematics of emergence.**Of what is becoming in the smallest slice of now. To differentiate is to
understand the logic of change— not in retrospect, but as it unfolds.
Because every moment contains momentum.
And the derivative captures that momentum—not by freezing it, but by framing it.
It does not describe where you've been. It reveals the direction you are becoming.

This is not just math for motion. It is math for transformation.
Wherever there is flow, there is a derivative.
In heat transfer, it becomes temperature flux. In finance, it becomes marginal return. In AI, it becomes the gradient—guiding machines to learn. In neuroscience, it maps the firing curve of a synapse over milliseconds.

To hold the derivative in your mind is to see not just a function, but a tendency. Not just a number, but a trajectory. Not just what is, but what wants to happen. It turns functions into narratives. Equations into unfolding decisions. Lines into lives in motion.

And perhaps most striking of all: the derivative lets us predict the future.
Not through mysticism— but by understanding that the current rate of change carries with it the next frame of reality. It tells us: If this is how the system bends—then this is where it will go.And that knowledge, zoomed to the edge of now, becomes the foundation for every field that dares to forecast.The derivative is not a tool. It is a window. And through it, we do not see stillness. We see the shape of becoming.It whispers something subtle but world-shaping:*Change is not an event. It is a gradient.*And to live within it—to model it, to adapt to it, to harness it—we must see not just the peaks of motion, but the way each instant leans into the next. The
derivative gives us the slope of becoming. But what about the total of all those becomings? What if we wanted not just to know the angle of the hill, but how much ground we've climbed? Not just the current, but the
accumulated flood? Then we must turn the lens in reverse. We must ask the inverse of change. And that leads us to the next idea— not about motion at a point, but motion as a sum. This is where we integrate.

5.3 The Integral: Slicing into Wholeness

Imagine trying to measure the area under a curve.
You can't just multiply height by width— the height is always changing.
The curve rises. Falls. Bends.
There is no rectangle that fits. No triangle that holds.
So what do you do?
You slice.

You break the area beneath the curve into narrow vertical strips. Each strip is thin enough to be almost a rectangle.
You measure each. You sum them all.
Then— you imagine slicing thinner. And thinner.
Until the width of each strip shrinks toward zero.
This is not a trick. This is a limit.

This is the integral:

$$\int_a^b f(x)\,dx$$

It is not a formula. It is a process:

- Cut the curve into infinitesimal pieces
- Measure each one
- Add them together
- Let the number of pieces approach infinity

You are not approximating. You are converging. Toward truth.

Where the derivative reveals what is changing, the integral reveals what accumulates.

The integral gives you:

- **Area** under a curve
- **Distance** traveled over time
- **Mass** from density
- **Work** from force over displacement
- **Total change** from rate of change

The world flows. The integral gathers that flow into meaningful sums.
Where the derivative zooms in, the integral zooms out.
One is local. The other is global.
One isolates. The other assembles.

And in a revelation that astonished the world— they were found to be perfect inverses.
The Fundamental Theorem of Calculus:

$$\frac{d}{dx}\left(\int_a^x f(t)\,dt\right) = f(x)$$

$$\int_a^b f'(x)\,dx = f(b) - f(a)$$

This theorem bridges the micro and the macro. It says:
What you accumulate through integration can be undone by differentiation.
And what you differentiate— you can sum back into meaning.

The curve that once resisted all analysis— can now be known. In every instant. And across any span.

This is calculus.

Not the study of change. But the study of how change becomes structure.

And in this structure, we find something profound:

The whole is not a mystery. It is a sum.

A sum of slivers. Of slants. Of slices of becoming.

Where algebra teaches symmetry and reversal, calculus teaches accumulation and consequence.

Every step taken, no matter how small, adds up. Every curve traced by time has a weight. Every breath, every signal, every ripple— can be measured if we are willing to listen finely enough.

To integrate is to gather what could not be held before. To restore continuity from fragmentation. To rebuild a total from the infinitesimal.

And more than that: It teaches us that what looks smooth and seamless is stitched together from a thousand invisible transitions.

Continuity is not simplicity. It is infinite coordination.

From physics to philosophy, from probability to engineering, integration tells us this:

Nothing flows without a cost. Nothing grows without accumulation. And nothing becomes whole without honoring what composes it.

5.4 Limits, Infinity, and the Edge of Definition

At the heart of calculus is an idea both simple and strange: you don't need to reach a point to understand what happens there. You only need to get close—infinitely close. This is the idea of the **limit**. It is not about arriving. It is about approaching.

$$\lim_{x \to a} f(x) = L$$

This notation means: as x gets arbitrarily close to a, the value of $f(x)$ gets arbitrarily close to L. No value is ever required to *equal* L. It only needs to get as close as you want—if you're willing to go far enough in.

Why does this matter? Because curves bend. Because rates change. Because reality doesn't always give you perfect values. You cannot always plug in a number. Sometimes it's undefined. Sometimes it's infinite. But even when the function breaks—its behavior can still be understood.

The limit is a window into what holds even when nothing else does. This is how we define the derivative: as a limit of ratios. This is how we define the integral: as a limit of sums. Everything in calculus relies on the edge. The edge of distance. The edge of time. The edge of knowability.

Infinity enters quietly. You break space into infinitely many pieces—infinitesimally small. You never hold them in your hand. But their total holds shape. It is paradoxical

and precise. Elusive and exact. Infinity is not a number. It is a direction. And calculus lives in that direction.
It uses infinity not to escape precision—but to arrive at it. This is why calculus works. Because even if the world never stays still, we can still understand what it's doing. By moving closer. By slicing finer. By allowing approximation to become certainty.

$$\lim_{n\to\infty} (\text{Approximation}_n) = \text{Truth}$$

This is not guessing. This is convergence. Limits are not boundaries. They are bridges. They show us: what lies on the horizon can still shape what we do now. Even if we never reach the edge,
we trace its outline with every step.

Limits whisper that proximity is power.
That the infinite is not unreachable—
only unholdable.

When you zoom in far enough,
every curve becomes a line.
Every motion becomes a slope.
Every accumulation becomes a form.

It is through the limit that motion becomes mathematics.
And the concept does not stay locked inside calculus.
It stretches outward.

- In series: how infinite steps can sum to a finite whole.
- In continuity: how unbroken change can emerge from points.
- In probability: how randomness converges into pattern.
- In physics: how the infinitesimal shapes the observable.
- In algorithms: how iteration finds fixed points.
- In thought: how understanding forms not from answers, but from approach.

The limit reframes certainty—
not as possession,
but as a direction of refinement.

$$\lim_{x \to a} f(x) = L$$

This is not a static fact.
It is a process frozen into symbol. Mathematics, through limits,
teaches us how to hold what can't be
held—
not by grabbing tighter,
but by letting approach be enough.

5.5 The World That Can Now Be Measured

Before calculus, the world was mostly described.
After calculus, the world could be predicted.

The stars were once divine. Their movements—mysterious.
But with calculus, Kepler's ellipses became Newton's laws.
And gravity— was no longer magic.
It was a differential equation.
Force = mass × acceleration

$$F = ma$$

Acceleration is change of velocity. Velocity is change of position. Each is a derivative of the one before.
Calculus turned motion into math. And math into foresight.

You could now know how fast a rocket must go to leave Earth.
How blood flows through arteries. How sound waves bounce off buildings. How populations grow. How diseases spread. How economies oscillate. How the brain adapts.
Wherever there is change— calculus listens.
It listens to rates. It listens to flow. It listens to hidden patterns beneath surface motion.

Differential equations now govern biology, chemistry, engineering, physics.
From the shape of a virus to the orbit of a satellite, from

the flicker of a neuron to the flicker of a star.
Calculus made the invisible visible— and the unknowable, knowable.

But beyond equations and predictions, calculus did something subtler.
It taught us that truth doesn't always come whole.
Sometimes, truth arrives in pieces.
In fragments. In limits. In convergence.
The world does not need to be flat for us to understand it.
It only needs to reveal how it moves.
And calculus was the tool we built to see that movement clearly. It gave us the mathematics of becoming. And with it, we began to see:
Reality is not static. It is unfolding.
Not in leaps, but in gradients. Not in snapshots, but in continuous change.

Calculus did not just measure motion. It mirrored the logic of transformation.

From the falling of an apple to the fusion in a star, from the way ink spreads in water to the way thoughts ripple across neurons— everything that changes carries a pattern.
And calculus gave us the grammar for that pattern.

It showed us that:
• Smoothness can hide infinite complexity • Growth obeys its own tempo • Systems do not shift arbitrarily— they evolve along curves • The smallest influence, if constant, will eventually reshape the whole

This was more than mathematics. It was a new kind of perception. A new kind of listening.
To integrate is to understand what builds. To differentiate is to understand what drives.

Together, they let us model not just what is— but what will be.
And for the first time, we weren't just observing the world. We were tracing its pulse.

Chapter 6

Probability – The Logic of Uncertainty

Not all truths are definite.
Some are measured in odds.
Some in risk.
Some in hope.
Probability
is the mathematics of perhaps.

6.1 The Coin That Never Promised You Anything

Flip a coin. What will it land on?
You don't know. You can't know.
But you can say: "There's a 50% chance it's heads."
This is not a guess. It's a structure.
It's what happens when we model the unknown.

Before probability, uncertainty felt like ignorance.
After probability, uncertainty became a field of study.
The coin becomes a system. Its outcomes become a space. Each possibility is mapped. Each chance is measured.
We call this the **sample space**:

$$S = \{\text{Heads}, \text{Tails}\}$$

And the probability of each outcome:

$$P(\text{Heads}) = \frac{1}{2}, \quad P(\text{Tails}) = \frac{1}{2}$$

This is not just gambling language. It is structure applied to uncertainty.

You don't know what will happen. But you know the shape of what could.
Probability quantifies ignorance— but transforms it into power.

This becomes profound in more complex systems.
Draw a card from a shuffled deck. Roll a die. Predict if it rains tomorrow. Assess the risk of disease. Calculate

the odds of an asteroid missing Earth.
In each case:

- You don't know the outcome.
- But you know the possible outcomes.
- And how often they tend to occur.

This is the core of probability.
Not clairvoyance. But calibration.
Not certainty. But reasoned expectation.

We say:

$$P(A) = \frac{\text{Number of favorable outcomes}}{\text{Total number of outcomes}}$$

But even more powerful than that is the concept of **conditional probability**:

$$P(A|B) = \frac{P(A \cap B)}{P(B)}$$

This asks:
What is the chance that A happens, given that B already has?
It is how doctors diagnose. How algorithms recommend. How weather forecasts shift.

The coin never promised you anything.
But probability does:
If you flip enough coins, they will obey the math.

$$\lim_{n\to\infty} \frac{\text{Number of Heads}}{n} = 0.5$$

This is the Law of Large Numbers.

It says randomness— when repeated enough— reveals pattern.
And that pattern is not an illusion. It is the shape of chaos seen from afar.

This is the paradox of probability: Each event feels unpredictable. But in aggregate, outcomes obey structure.
The uncertain becomes reliable through repetition. The scattered becomes stable through scale.

And from this, a deeper principle emerges:
The Central Limit Theorem.
It says: no matter the original distribution, the sum of many small, independent events tends toward a bell curve— the normal distribution.

It is why human heights cluster. Why test scores follow patterns. Why noise, when averaged, becomes signal.
The randomness doesn't disappear. It self-organizes. It converges.

Probability, then, is not about control. It's about tuning. You cannot dictate the outcome of a single coin. But you can build systems that anticipate the shape of many.
Insurance. Forecasts. Machine learning. All of them rest on this strange wisdom:
Uncertainty is not the enemy of understanding. It is its foundation.

The coin never promised you certainty. But mathematics did promise you meaning. And when you listen across enough flips, meaning appears.

6.2 Bayes' Rule: Updating What You Believe

In life, you rarely start from scratch.
You begin with a belief— a hunch, a suspicion, a model.
Then something happens. New data arrives. A test result. A weather report. A headline.
And the question becomes:
How should this change what I believe?

This is the question Bayes' Rule answers.
It is a formula for revision. Not of the world— but of your mind.

$$P(H|E) = \frac{P(E|H) \cdot P(H)}{P(E)}$$

Where:

- $P(H|E)$ is the **posterior**: the updated belief
- $P(H)$ is the **prior**: the belief before evidence
- $P(E|H)$ is the **likelihood**: how expected the evidence is under the hypothesis
- $P(E)$ is the **normalizing factor**: how likely the evidence is in general

Bayes' Rule doesn't predict the future. It predicts how your beliefs should shift when the future becomes present.

Example: You take a medical test.

- The disease is rare: 1 in 1000 people have it.

- The test is 99% accurate.
- Your test comes back positive.

Should you panic?
Intuitively, yes.
But probabilistically?
Let's compute:

- $P(\text{Disease}) = 0.001$
- $P(\text{Positive}|\text{Disease}) = 0.99$
- $P(\text{Positive}|\text{No Disease}) = 0.01$
- $P(\text{No Disease}) = 0.999$

Using Bayes:

$$P(\text{Disease}|\text{Positive}) = \frac{0.99 \cdot 0.001}{(0.99 \cdot 0.001) + (0.01 \cdot 0.999)} \approx 0.09$$

So even after a positive test, your actual chance of having the disease is about 9%. Why?Because the disease is rare, and false positives matter. **Bayes doesn't give you truth. It gives you adjustment.** It teaches humility.Bayes' Rule isn't just for doctors. It's used by:

- Self-driving cars
- Spam filters
- Machine learning algorithms
- Financial models

- Criminal forensics
- Weather forecasts

But more than all of that— Bayes is a way to live. **Start with a belief. Stay open. Let new evidence refine you.** This is the logic of learning. This is the math of rational change.Bayes teaches that knowledge is never final. Only provisional. And if your priors never shift, your mind isn't holding fast— it's standing still. Every moment is a new data point. Every experience—an update. The question is not whether you're certain. But whether you're adjusting in proportion to the surprise. Bayes reminds us:

- Belief is not binary.
- Confidence is not conviction.
- Certainty should always be conditional.

It's not just the test result that matters. It's how expected that result was, given your assumption. And how rare it would be if you were wrong.This isn't just a formula. It's a philosophy:

> Truth is not what you start with. It's what you converge toward.

Bayes is the map we use when facts arrive piece by piece. When certainty is fogged. When decisions must still be made.And it says: You don't need to know everything. You only need to adjust precisely. And keep adjusting—until the shape of reality emerges from the noise.

6.3 Entropy and the Measure of Surprise

Uncertainty is not always equal.
Flip a fair coin? That's full uncertainty.
Flip a coin weighted to land heads 99% of the time? That's almost certainty— even though two outcomes still exist.

So we ask:
Can uncertainty be measured?
The answer is yes.
And the measure is called **entropy**.

Developed by Claude Shannon in the 1940s to formalize communication, entropy became the heartbeat of information theory.
It is defined as:

$$H(X) = -\sum_i P(x_i) \log_2 P(x_i)$$

Where X is a random variable and x_i are its possible outcomes.

What does this mean?
Each outcome carries a **surprise**:

$$\text{Surprise of } x_i = \log_2 \left(\frac{1}{P(x_i)} \right)$$

Rare things are more surprising. Common things, less so.
Entropy is the **average surprise** you should expect from

a distribution.

If every outcome is equally likely, entropy is maximized. You are fully in the dark.
If one outcome dominates, entropy shrinks. The system becomes predictable.

Example:

- Fair coin: $P(\text{Heads}) = 0.5$, $P(\text{Tails}) = 0.5$

$$H = -[0.5 \log_2 0.5 + 0.5 \log_2 0.5] = 1 \text{ bit}$$

- Biased coin: $P(\text{Heads}) = 0.99$, $P(\text{Tails}) = 0.01$

$$H \approx 0.08 \text{ bits}$$

So even though both coins have two outcomes, only one is truly uncertain.

Entropy now guides:

- Data compression (what's worth storing?)
- Machine learning (what features hold surprise?)
- Cryptography (how predictable is the key?)
- Physics (how ordered is a system?)
- Neuroscience (how unexpected is this input?)
- Decision theory (what reduces the most uncertainty?)

Entropy doesn't measure chaos. It measures possibility.

How many ways could this be? In a low-entropy world, surprise is rare. In a high-entropy world, it's the norm. Entropy is the system's confession about what it refuses to predict. And in every field that seeks to know— from biology to AI— entropy remains the question: **How much don't you know?** Entropy does not ask what is true. It asks what is possible— and how spread out that possibility is. In thermodynamics, it measures disorder. In information theory, it measures choice. In both, it whispers the same idea:

> The more ways a system can be arranged, the less certain you are about what you'll find.

Entropy isn't noise. It's freedom.The freedom of a system to take many forms. The freedom of a message to
contain surprise. The freedom of nature to explore without a script. But freedom isn't free. In computation, high entropy means more bits to transmit. In life, more entropy means more room for error— but also more room for creativity. That's why entropy lives at the core of learning. Because surprise is the signal. It's what tells you: "Something's new here." "Your model needs to stretch." When a brain updates, when an algorithm adjusts, when a theory evolves— it's not fighting entropy. It's listening to it.Entropy doesn't just measure uncertainty. It measures potential. And in every system that learns, that adapts, that encodes, decodes, and evolves— entropy is the pulse that guides the change.To know entropy is to embrace a truth:Knowledge is not the absence of uncertainty— but the dance with it.

6.4 Randomness, Pattern, and the Shape of Noise

Some systems look random only because we haven't found the pattern yet.
Others truly are.
But how do you know which is which?
This is the central puzzle of every scientist, statistician, and signal analyst.

Noise is variability without pattern. **Signal** is pattern within variability.
Your task is to distinguish the two.

When you stare at static on a screen, your brain searches for shape. It's built to.
We are *pattern-detection machines.*
But that strength is also a vulnerability.
We see meaning where there is none.
We find faces in clouds. Voices in noise. Trends in randomness.
This is called **apophenia**— false pattern recognition. So probability asks: What patterns are real? It does this by modeling noise. *White noise*: equal power at all frequencies. It is structureless. Its entropy is maximal. It carries no compressible pattern. If something behaves like white noise, you cannot predict it. You can only describe its distribution. But if there's **autocorrelation**— a tendency for past values to influence future ones— then something else is present: **memory.**

structure. a signal. This is how we find meaning in data. We measure:

- **Variance** — how far do things spread?
- **Covariance** — do two things vary together?
- **Fourier spectra** — which frequencies dominate?
- **Mutual information** — how much does one variable tell us about another?

And if enough structure is found— we build a model. If no structure is found— we model the randomness itself. **Not all noise is useless.** Some noise contains clues. **Stochastic resonance** shows that a small signal becomes more detectable with the right amount of noise. Too little? You miss the pattern. Too much? You drown it out. Just enough? Clarity.

This tells us something deep:
Randomness is not the opposite of meaning. It is the background from which meaning emerges.
Noise is not failure. It is the stage where pattern might appear. The randomness you see is not chaos. It is potential without bias. A space where nothing is favored—yet anything might emerge.

But pattern? Pattern is asymmetry. Pattern is deviation from the void. Pattern is what refuses to vanish when the averages are stripped away.

And so we study randomness not to eliminate it— but to reveal the substrate against which structure becomes legible.

There are entire fields—like statistical mechanics— built on the premise that noise, when viewed at scale, obeys laws. That temperature, pressure, and entropy all rise from the collective unpredictability of individual particles.

In finance, randomness models risk. In neuroscience, it explains synaptic variability. In quantum physics, it isn't a bug—it is the feature.

And yet, even randomness has shape. Not everything unpredictable is equally wild.

- **Uniform noise** spreads possibility evenly.
- **Gaussian noise** centers around a mean.
- **Pink noise** decays with frequency—found in heartbeats, music, and nature.
- **Brownian noise** accumulates—like stock prices or pollen drifting in water.

Each flavor of noise tells a different story. Each one is a fingerprint of how variability behaves in a given domain.

Even randomness has degrees. Even chaos whispers structure—if only faintly.

That's why we build models not just to predict— but to delimit where prediction fails. To say: "Here is what we know. Here is how far the signal reaches. And here, beyond that—lies the void."

But this void is not empty. It is filled with fluctuation, suggestion, emergence.

A single neuron spikes erratically— but millions together produce perception.
A stock price zigzags unpredictably— but portfolios curve toward equilibrium.

This is the hidden lesson:

> The more you understand randomness, the more you see that meaning is not a sharp line. It is a gradient— blurring out from certainty into possibility, from clarity into noise.

And perhaps that is where all structure begins: At the threshold of what we can't yet explain— but dare to model anyway.

6.5 Probability as Strategy: Betting, Decisions, and Belief

Probability isn't just about the world.
It's about how you move through it.
It's not a description. It's a strategy.

Imagine a game:
You draw from a deck of cards. If it's red, you win \$10.
If it's black, you lose \$5.
Would you play?
Most people ask: "What's the chance I'll win?"
But the deeper question is: "What is this gamble worth?"
Expected value answers that.

$$EV = (P_{\text{win}} \times \text{gain}) + (P_{\text{lose}} \times \text{loss})$$

$$EV = (0.5 \times 10) + (0.5 \times -5) = 5 - 2.5 = 2.5$$

The game is in your favor. You should play.
Even if you lose once, you'll win in the long run.
This is probability as decision.

But the world isn't always a casino. Sometimes, stakes are hidden. Outcomes are unclear.
Probabilities are personal.
Enter subjective probability.
This is where Bayesian thinking lives— where your belief about what is likely guides your actions.

In the courtroom: jurors weigh evidence. They don't get

certainty. They get plausibility.
In medicine: a doctor doesn't know the disease with certainty— she updates her diagnosis based on test results and prior cases.
In life: you choose a partner, a job, a home with incomplete information.
But you choose anyway.

That's what probability becomes:
A philosophy of incomplete knowledge.
A logic for uncertain action.
It helps you:

- Know when to wait
- Know when to risk
- Know what to update
- Know how to protect

Game theory grows from this.
So do:

- Investment strategies
- AI decision trees
- Insurance models
- Risk management
- Evolutionary biology

They all rely on this truth:
You don't need to know what will happen. You only need to respond wisely to what might.

So probability, in the end, isn't just about events.
It's about choices.
It's how you move forward when certainty won't come.
This is why probability matters more than prediction. Because it doesn't just model the world. It models you—as a thinker, a decider, a learner in flux.

It gives you the language of caution without fear, and risk without recklessness.

You don't flip coins in real life. You weigh options with imperfect data. You act under fog.

Probability becomes your compass when the map is blurry.

- It tells the pilot how likely turbulence is.
- It tells the trader when the market is overheating.
- It tells the parent what the odds are that the fever is nothing—or something more.

It tells you how to move—not because you know, but because you've calibrated what you don't.

And that's where expected value becomes expected utility.
Not all outcomes are equal. Losing 100$whenyouhave$1,000 stings. Losing 100$whenyouhave$105 destroys you.

So now we ask: not just, "What will I win?" But: "What does it cost me to lose?"

Utility curves emerge. Rational agents become risk-averse, or risk-seeking, depending on where they stand on the curve.

This is no longer theory. It's how we design AI agents that drive cars, robots that explore Mars, and systems that detect fraud, disease, and catastrophe.

Because at its core, probability isn't math. It's adaptation. It's how systems—biological or digital—act when nothing guarantees a result.

So whether you're placing bets, making policies, or just deciding what comes next—
Probability does not offer certainty. It offers resilience.

> A way to choose—
> when the ground won't settle.

6.6 Limits, Infinity, and the Edge of Definition

At the heart of calculus is an idea both simple and strange: you don't need to reach a point to understand what happens there. You only need to get close—infinitely close. This is the idea of the **limit**. It is not about arriving. It is about approaching. It is the mathematics of
getting closer and closer without ever having to be there. A way of understanding a destination by traveling toward it forever.

$$\lim_{x \to a} f(x) = L$$

This notation tells us: as x moves ever closer to a, the value of the function $f(x)$ moves ever closer to some value L. Not exactly. Not always. But close enough—for any precision we choose. No value is ever required to equal L. It only needs to get as close as we wish, if we're willing to go far enough along the approach.

This is more than mathematical notation. It is a philosophy. A way of thinking about change when exactness is elusive. Limits allow us to grasp the behavior of systems at their edges—where the surface cracks, where definitions blur, where ordinary math would throw up its hands. Limits step in where numbers cannot go.

Why does this matter?

Because curves bend. Because rates vary. Because in

reality, things don't always behave in neat, discrete steps. Instead, everything is motion. Everything is shifting. And in such a world, limits give us the ability to hold onto certainty by letting go of finality.

You cannot always plug in a number. Sometimes the function is undefined. Sometimes it shoots toward infinity. Sometimes it breaks entirely. But even when a value doesn't exist at a point, we can still describe what's happening around that point. We can still understand its shape.

The limit is a window into what holds even when nothing else does.

This is not just useful. It is revolutionary. This concept powers every idea in calculus.

The derivative, for example, is defined as the limit of a ratio. Not just any ratio, but the ratio of change. The speed of a curve. The slope of a line that never quite touches the curve, but leans into it tighter and tighter:

$$f'(x) = \lim_{h \to 0} \frac{f(x+h) - f(x)}{h}$$

This tiny expression holds the secret of motion. It is how we measure velocity at a single instant. How we define acceleration. How we understand dynamic systems where everything moves, but we still want to make sense of it all.

And then comes the other half of calculus: the integral. The sum of infinitely many infinitely small pieces. A way to measure the area under a curve, even when that area resists neat rectangles. Even when it bends and warps and tapers toward infinity. We don't calculate it directly.

We let limits do the work.

$$\int_a^b f(x)\,dx = \lim_{n\to\infty} \sum_{i=1}^{n} f(x_i)\Delta x$$

We do not merely estimate. We let the approximations converge into truth.

This is how limits function: not by arriving, but by revealing. Not by forcing closure, but by welcoming precision through endless approximation. Limits ask us to trust that the truth will emerge—not in a single step, but in the infinite refinement of steps.

And this brings us to one of the most powerful, and strangest, concepts in all of math: **infinity**.

Infinity is not a number. You cannot write it on a receipt. You cannot reach it, hold it, or compare it in any normal sense. But it is real—mathematically real—in the same way that direction is real. It points the way. It stretches the line. It tells us: this process will not stop.

Infinity is how we break the world into pieces too small to grasp, and then reassemble them into something that makes sense. It is how we define the area of a circle. The curve of a spiral. The wave of a sound. The shape of a probability.

It is paradoxical and precise. Elusive and exact. And calculus lives in that paradox.

We do not use infinity to avoid precision. We use it to reach it.

We define a curve by what it becomes, not just what it is. We define a behavior by what it tends toward, not only what it equals. This is not guessing. This is convergence.

$$\lim_{n\to\infty}(\text{Approximation}_n) = \text{Truth}$$

It is the fundamental act of mathematics: taking what is incomplete, and—through structure, patience, and reasoning—turning it into certainty.And this is where limits shine most brightly: at the edge.At the boundary of what is defined and undefined. At the place where a function might not exist, but its behavior still speaks. At the place where our knowledge seems to falter—but math continues. Limits are not boundaries. They are bridges. They are how we cross from known to unknown. From discrete to continuous. From the visible to the infinitely small.And in this way, the limit is more than a tool. It is a philosophy of learning.To move closer. To zoom in. To refine. To understand something not because we arrived, but because we committed to approach. This is how truth is found at the edge of definition. Not with finality— but with infinite attention.

Chapter 7

Geometry – The Logic of Space

> *"Geometry is knowledge of the eternally existent."* — Plato

Before symbols, before equations, before algebra taught us to
generalize and calculus taught us to change—there was space. And there were those who needed to understand it. To rebuild fields after floods. To align temples with stars. To carve meaning from horizon lines. Geometry was not born in abstraction. It was born from land and light, from shadow and span, from the body's need to anchor itself in a moving world.

Space was not empty. It was structured. And geometry became its grammar

7.1 From Land and Shadow: The Birth of Geometry

Long before equations, there was rope.
Tied in knots, stretched tight— used by Egyptian "rope-stretchers," to remeasure farmland after each Nile flood.
Geometry was practical before it was theoretical. It was not invented to beautify textbooks. It was born from survival. From territory. From the need to restore what nature had erased.

To rebuild borders, they needed to understand angles, lengths, and areas.
They measured. They marked. They stretched triangles across the earth.
What they discovered— almost unknowingly— was that space has rules.
The triangle that always adds to 180°. The circle that encloses the most area. The square that forms perfect right angles.
These weren't human inventions. They were structures already present in the world— discovered through the quiet persistence of measurement.

In Babylon, scribes used clay tablets to record the lengths of sides and diagonals— the earliest whispers of the Pythagorean theorem etched into earth.
In India and China, astronomers built observatories to track the heavens with angular precision. Stars became coordinates. Time became geometry.

Even without proving it, they knew something profound:
Space could be understood.

But in Greece, a shift happened.
A revolution.
What if geometry didn't just describe space— but proved it?
Enter Euclid. 300 BCE. Alexandria.
His *Elements* wasn't just a textbook— it was a new architecture of thought. A cathedral built from postulates.
Five simple assumptions. No more. And from them— an entire universe of form.

He began not with numbers, but with definitions:

- A point is that which has no part.
- A line is breadthless length.
- A straight line lies evenly with the points on itself.

From these, came hundreds of theorems. Each built from logic. Each stacked like stone. Euclid showed the world that geometry wasn't about guessing— it was about knowing. **Geometry became philosophy.** To prove a triangle was to prove a truth about the world itself. And so, to know the shape of space was to know the shape of truth. But the story does not end in Alexandria. Across centuries, geometry kept evolving. In the Islamic Golden Age, mathematicians translated and expanded Euclid— inventing new techniques for area, optics, and architecture.In the Renaissance, artists rediscovered perspective— turning geometry into vision. And in the modern age, Gauss, Lobachevsky, and Riemann bent geometry itself— imagining curved spaces where parallel lines might meet. And all of this— all of it— began with a rope pulled tight across the floodplain. A tool. A line. A truth.

Geometry was not born in abstraction. It was born from land and shadow. From bodies needing borders. From eyes chasing light. From hands drawing paths across the unknown.It began with the earth. It reached toward the stars.And it built the bridge between them.**This is the logic of space.** Not cold.Not distant.But ancient.Intimate. Enduring.

7.2 Euclidean Order: Proof, Axioms, and the Infinite Flat Plane

What makes a truth eternal?
For Euclid, the answer was: *Start from what cannot be doubted. Then build what cannot be denied.*
He didn't rely on authority. He relied on clarity.

He began with five postulates— simple, unprovable, yet seemingly self-evident:

1. A straight line can be drawn from any point to any other.
2. A finite straight line can be extended indefinitely.
3. A circle can be drawn with any center and radius.
4. All right angles are equal.
5. If a line intersects two others and creates interior angles less than two right angles, those lines will eventually meet on that side.

These were not theorems. They were foundations.
From them, he built the first logical universe.

Definition. Axiom. Theorem. Proof.
Geometry became a machine. Feed in assumptions, and out came certainties.
You didn't need to see or measure. You could reason.
This was a radical shift.

No longer did a truth rely on sight. It relied on sequence.
A triangle didn't need a ruler. It needed a reason.

And once that reason was proven, it held—forever.

Euclid's geometry defined the flat world: the infinite plane, the straight edge, the perfect corner.
From these five postulates, you could prove:

- The sum of interior angles of a triangle is 180°
- Opposite angles are equal
- Parallel lines never meet
- Similar figures preserve shape across scale
- The area of a circle depends only on radius

Each theorem was a truth not just about drawing, but about possibility.

And more than that— it was a truth about truth.
Geometry became a map of reason itself.
Every proof became a path. Every theorem,
a destination. Every construction, a sentence in logic.And once proven, you never had to doubt it again. This was not mathematics as calculation.

It was mathematics as
revelation. In the Euclidean world, every shape obeys. Every line knows its role. Every corner tells the same story. Even infinity became manageable. You could extend a line
forever— not because you ever reached the end, but because the postulate told you it had no end.This was not
speculation. This was structure.But perhaps Euclid's greatest gift was not any single result— but the method itself.The architecture of reasoning. The idea that truth can be built. Brick by brick. Step by step.And this idea outlived him.**It shaped science. It structured philosophy. It became the blueprint of every proof ever written.**Because of Euclid, we don't just ask what is true. We ask: *What must be true— if this is?* That is the Euclidean world. Not just flat space— but flat logic.A surface where everything unfolds without contradiction,as long as the axioms hold.This is the elegance of the infinite plane: Not chaos. Not randomness. But pure, deductive design.Euclid didn't just
describe
geometry. He defined what it meant to know anything at all.

7.3 The Crisis of the Fifth Postulate: Birth of Non-Euclidean Worlds

For centuries, mathematicians trusted Euclid's vision. But one postulate stood apart. The fifth. **If a line intersects two others and forms interior angles less than two right angles, those lines will eventually meet.** It felt clumsy. Less obvious. More conditional. Unlike the others, it wasn't purely local. It made a claim about the distant future of lines. Mathematicians tried to derive it from the first four. Failed. Tried again. Failed. For two thousand years. Then came a different idea: **What if the fifth postulate is not necessary? What if the universe does not obey it?** In the early 1800s, Lobachevsky in Russia, Bolyai in Hungary, and later Riemann in Germany all explored geometries where the fifth postulate was modified or removed. What they found shattered the map. **In hyperbolic geometry:** Parallel lines diverge. Triangles add to less than 180°. Space curves away from itself. **In spherical geometry:** There are no parallel lines. Triangles add to more than 180°. Space curves toward itself. The flat plane, once sacred, was just one possibility. **Geometry was no longer absolute.** It was conditional. Dependent on the shape of the space itself. And suddenly— geometry was not just about space. It was about the nature of the universe. Curved geometries

weren't just logical toys. They were keys to understanding gravity, light, motion. They would one day become Einstein's spacetime. **The rejection of the fifth postulate opened the door to the shape of reality itself.** It was no longer enough to ask, "What is geometry?" The question became: "Whose geometry?" The universe, it turned out, was not beholden to our intuition. What we assumed as flat—was only approximately so. What we called straight—was bent by mass and motion. Einstein would later prove it: Gravity is not a force. It is geometry. Matter tells space how to curve. Space tells matter how to move. And suddenly, the fifth
postulate—so awkward, so
debated— became a whisper from the edge of physics. A signal that the structure we live in may not be the one we once drew. This crisis birthed a liberation.
Mathematics was no longer a mirror of flat surfaces. It became a canvas for all conceivable shapes. **Non-Euclidean geometry** is not a rebellion. It is a revelation. That truths can coexist, even when they contradict what we first
believed. The real breakthrough was not that space could curve— but that thought could, too. Mathematics, like the universe it seeks to describe,
became curved from the inside out.

7.4 Curved Spaces, Bent Realities: Geometry After Euclid

Once forbidden, curved geometry
became the bridge between abstraction and reality.

In 1854, Bernhard Riemann stood before the faculty at Göttingen and proposed a new geometry.
Not as a rejection of Euclid— but as a generalization.
A space could be flat, but it could also curve.
And that curvature could be measured with intrinsic tools— from within the space itself.

No longer was geometry a property of diagrams on paper.
It became a property of the universe.
A triangle drawn on a sphere has angles that sum to more than 180°.
A triangle drawn in hyperbolic space sums to less.
The angle sum reveals the curvature of the space beneath.

Einstein took this insight and made it physical.
In 1915, his general theory of relativity replaced gravity with geometry.
Mass curves space. Light follows that curve.
The shortest path is no longer a straight line in the Euclidean sense.
It is a geodesic— a line drawn by the space itself.

This isn't metaphor.
Starlight bends. Time dilates. Orbits precess.
And we observe it all.

Satellites, clocks, black holes— they all obey curved geometry.

What began with ancient rope now guides spacecraft through a bent cosmos.
What began with idealized triangles now models the expansion of the universe.
Geometry is no longer static. It flows. It flexes. It responds.
Euclid gave us rules. Riemann gave us freedom.
The shape of space is not assumed. It is discovered.
And what we discover— is not uniformity, but variance. Space is not the same everywhere. It stretches near stars. It collapses near black holes. It ripples with the echoes of ancient collisions.

Gravitational waves—predicted by Einstein, confirmed a century later— are tremors in the geometry itself. Space quivers. Time shifts. And we have built machines sensitive enough to feel it.

What used to be myth— that the heavens react to gods or giants— has become mathematics. And the equations do not blink.

We once thought geometry was the study of shapes. But now we see: It is the study of the possible. Of how paths bend. Of how time warps. Of how existence can be stitched differently, from point to point, depending on what inhabits it.

A flat space is just one note. The cosmos plays a full spectrum.

And still— we are only mapping what the universe allows. There may be curvatures we've never imagined. Dimensions folded too tightly to see. Manifolds that host realities we do not yet inhabit.

Riemann opened the gate. Einstein walked through. Now we send our telescopes, our satellites, our minds— to follow.

Geometry is no longer the background. It is the actor. And the stage. And the script.

In curved space, reality doesn't sit still. It evolves. And we are learning how to read its motion.

7.5 Topology: When Shape Ignores Distance

Geometry asks: How big is the angle? How far is the edge?
Topology asks: What stays connected no matter how you bend it?

If geometry is the art of measurement, topology is the art of structure without scale.
A coffee mug and a donut are the same.
Both have one hole. Both can be stretched into each other without cutting or gluing.
In topology, distance dissolves. Only continuity remains.

Imagine a world made of rubber.
You can stretch, twist, fold— but you can't tear or fuse.
A square becomes a circle. A line becomes a loop. A loop becomes a sphere.
What matters is not shape, but essence.
How many holes? How many surfaces? Is it one piece or many?

Topology uncovers the hidden skeleton of form.

- A Möbius strip— one side, one edge, no inside or out.
- A torus— the donut space, where paths can loop endlessly.
- A Klein bottle— a shape that folds into itself, uncontainable in ordinary space.

This shift—from measuring to mapping— changed how we understood not only geometry, but reality itself.
Topology became essential in:

- Classifying particles in quantum fields
- Modeling networks in data science
- Understanding the shape of the universe

Even DNA— with its knots and loops— obeys topological laws.

Topology revealed that some truths are deeper than dimensions. Some identities cannot be seen by eye, but only traced through continuity.
It is not the distance that defines the object— but the logic of how it holds together.

Chapter 8

Logic – The Shape of Truth

"Reason is the line that holds when the world breaks apart."

8.1 The Axiom and the Argument

Before there were numbers, there were reasons.
Before there was algebra, there was the syllogism.
Long before equations or computers or physics—
there was a desire to know what must be true.

Logic begins not with symbols, but with trust.
Trust that a certain sequence of
thoughts leads unavoidably to a conclusion.

All humans are mortal. Socrates is human.
Therefore, Socrates is mortal.
This is not opinion. This is structure.
It is the shape of reason laid bare.

The Greeks studied it. Aristotle named it.
He called it syllogism: a way to combine premises to forge a conclusion.
It wasn't just argument. It was architecture.

Each premise a pillar. Each rule of inference a beam. Each conclusion a ceiling that must hold if the base is sound.

But the Greeks did not formalize it all.
Their logic was still natural— linguistic, not yet symbolic.
It took two thousand years for logic to become mathematics.
That turning point came when mathematicians asked a deeper question:
What is the foundation beneath every truth?
Their answer: Axioms.

Assumptions so basic, they are not proven— only accepted.
From them, all other truths would be derived.
Like geometry, but for thought itself.

Logic became a discipline. Not the art of persuasion— but the science of deduction.
And every argument became a system.
If the premises are true— and the rules are sound— then the conclusion is guaranteed.
No faith. No opinion. No authority.
Just structure.
And once logic had structure— it could be studied like any other object. It could be built. Or broken.

From this, formal systems were born. Rules to generate proofs. Symbols to remove ambiguity. Languages so precise, they could host certainty itself.

Peano axioms for arithmetic. Hilbert's axioms for geometry. Zermelo-Fraenkel for sets. Each system was a cathedral— erected not of stone, but of statements.

But then came Gödel.

In 1931, Kurt Gödel showed something devastating and beautiful: Even the most carefully built system— if powerful enough to express arithmetic— contains true statements that cannot be proven within it.

This is Gödel's incompleteness theorem.

It shattered the dream of a final foundation. Not because logic failed— but because its reach is always larger than its grasp.

There are truths we can write, that no ladder of inference can reach.

And yet— this was not defeat.

It was liberation.

Because it told us that mathematics is not a closed box. It is an open terrain. A landscape of consequence— not dictated by consistency alone, but by the axioms we choose to begin with.

Change the axioms— and you change the world they define.

Euclidean or Riemannian. Classical or quantum. Finite or infinite.

Every system is a mirror of its starting points. Every argument, a path through its own terrain.

Logic is not the end of philosophy. It is its instrument. Its scaffold. Its compass.

And an axiom— that quiet assumption at the beginning of it all— is not the enemy of truth.

It is the invitation to build one.

8.2 Truth Tables and Boolean Bones

What if every statement could be either true or false—nothing in between?
And what if you could write rules that operated on truth itself like numbers?

Enter George Boole.
In the 1800s, he transformed logic from verbal argument into algebraic structure.
Let `true` = 1, `false` = 0.
Now logic could be calculated.

- **AND** becomes multiplication: $1 \cdot 1 = 1$ (true), $1 \cdot 0 = 0$ (false)
- **OR** becomes addition (with limit): $1 + 0 = 1$, $1 + 1 = 1$
- **NOT** becomes negation: $\texttt{NOT}(1) = 0$, $\texttt{NOT}(0) = 1$

Logic now had a skeleton.
Each operation had a table. Each table a function. Each function a pathway for thought.

Truth tables became maps of reasoning:

A	B	A AND B
0	0	0
0	1	0
1	0	0
1	1	1

Boole never saw a computer.
But his logic became their bones.
Every digital circuit— every decision in your phone, your laptop, your AI— rests on Boolean logic.

Logic gates are physical manifestations of truth tables.
AND, OR, NOT become metal switches.
Semiconductors wired to mimic meaning.

With Boolean algebra, logic no longer lives in minds. It lives in machines.
Thought became hardware. Truth became electricity.
And from these humble tables rose a global computation engine. But an engine of what kind?

An engine not of pistons or combustion— but of decision.
An engine that runs on possibilities.

Because once logic is made of 1s and 0s, you can combine it, layer it, scale it.

From a single gate, you build adders. From adders, processors. From processors, minds that count, compare, predict.

It starts with a simple proposition: Is it raining? But Boolean logic lets you build from that to: Should I bring an umbrella if it's raining AND I'm walking?

From weather apps to rocket guidance, from medical scanners to video games— this is not "intelligence." This is logic, scaled.

Truth tables are not quaint tools of old philosophers. They are blueprints for modern architecture.

• If this, then that. • If both, then proceed. • If neither, then halt.

Every algorithm you use— every pixel that renders— is the echo of Boole's insight.

And the real leap was not just in encoding logic— but in making it programmable.

Because once logic can be modular, it can be abstract. It can generalize.

You are no longer writing outcomes. You are designing systems that decide on their own.

Boolean algebra is not a relic. It is the backbone of interactivity.

And every complex machine— every smart assistant, search engine, digital dream— still hums with the binary beat of 1 and 0.

Boole gave us bones. We wrapped them in circuits. Then taught them how to move.

8.3 Proving with Symbols: The Rise of Formal Logic

Once logic could be computed, the next ambition was larger:

Can all of mathematics be built from logic? Can every truth be proven, step by step, from a perfect system of rules?

This dream lit the minds of 19th-century mathematicians.

Frege led the way.

He replaced words with symbols. Sentences became formulas. Reasoning became syntax.

No more "Socrates is mortal." Now: $M(s)$

If all x are mortal: $\forall x\ M(x)$

And s is human: $H(s)$

Then with the right axioms, you derive: $M(s)$

These weren't just fancy notations. They were precise blueprints.

Logic became programmable.

Each line followed from the last by strict rules of inference.

No vagueness. No leaps. Only deduction.

David Hilbert expanded the dream.

He imagined all mathematics as a single complete formal system.

Start with axioms. Add inference rules. Build everything. Prove everything.

In this world, truth wasn't discovered— it was constructed.

Like a cathedral of reason, mathematics would rest on foundations that could not crack.

But this vision would not last.

Because just when the dream seemed possible— a deeper truth appeared.

Not all truths can be proven. Some structures, if consistent, must be incomplete.

This revelation did not come from outside. It came from within logic itself.

And it came with a name:

Gödel. Not with contradiction, but with construction.

He didn't break the system. He built a statement it couldn't reach.

A sentence that looked inward and asked: "Am I provable within this world?"

And suddenly, the mirror cracked.

If the system could prove the sentence, it would be false.

If the system could not prove the sentence, it would be true— but unprovable.

Either way, something had slipped through the net.

Gödel had found the edge of formal logic. And he drew a clean line around it.

No matter how flawless the rules, how careful the axioms, any system powerful enough to speak arithmetic must contain truths it cannot prove.

Truth, it turned out, was not the same as provability.

Hilbert's cathedral still stood—but now with shadowed alcoves. Rooms that could not be lit from within.

And deeper still: No consistent system can even prove its own consistency. A structure cannot carry the weight of its own self-trust.

The ambition had been totality. But what emerged was incompleteness. Not as a failure— but as a law of structure itself.

From that moment forward, logic was no longer a ladder to omniscience. It was a landscape with boundaries.

A system could be vast, elegant, powerful— but never whole.

Gödel's work didn't end formalism. It refined it. It taught mathematics what it could never forget again:

That to speak all truth, you sometimes must leave the system that speaks.

And just beyond Gödel's theorem stood another threshold.

Another mind.

Another impossibility.

Turing.

8.4 Gödel's Crack in the Foundation

In 1931, a quiet, brilliant mathematician named Kurt Gödel proved that the dream of total certainty was impossible.

He took the very tools of formal logic— symbols, axioms, inference rules— and used them against themselves.
His question: Can a system prove everything true within it, without contradiction?
Hilbert had hoped the answer was yes. Gödel proved the answer was no.

First Incompleteness Theorem: In any formal system powerful enough to express arithmetic, there exist true statements that cannot be proven within the system.
Second Incompleteness Theorem: Such a system cannot prove its own consistency.

The implications were seismic.
Not just for logic. But for mathematics. And for the philosophy of truth.

If your system is strong enough to express $1 + 1 = 2$ and beyond, then somewhere inside there will be a sentence that says:
"This statement cannot be proven."
And if it can be proven— the system is inconsistent.
If it cannot be proven— it is incomplete.
Either way, certainty breaks.

Gödel encoded logic within logic.

He gave numbers to statements, statements to proofs, and allowed math to speak about itself.
This self-reference— like a snake swallowing its tail— revealed something new:
There is no perfect mirror.
No final system that contains all truth without tearing at the edges.

Logic, like language, contains its own limits.
And those limits are not errors. They are necessary.

Yet within this fracture is a deeper beauty.
The universe of truth is larger than any system we can build.
And so, even in logic, mystery remains.

8.5 Logic Beyond Logic: Paradox, Incompleteness, and Infinity

Even before Gödel, logic had already begun to tremble.

In 1901, Bertrand Russell found a problem at the heart of set theory.
The set of all sets that do not contain themselves—does it contain itself?

If it does—then by definition, it shouldn't.
If it doesn't—then it should.
This is Russell's Paradox.
A hole in the foundation where contradiction loops endlessly.

Set theory, once seen as the bedrock of all mathematics, now had a crack.
Naive logic had to be rewritten.
Zermelo and Fraenkel stepped in. They rebuilt set theory with new rules— banning certain kinds of self-reference. Mathematics would continue— but the innocence was gone.

Then came Gödel. Then came Turing. And with them, the revelation that logic and computation share the same boundary.

Turing showed that some problems are not just hard— they are undecidable.

No machine, no algorithm, can solve them in general. We learned that truth, proof, and process are not the same.Some truths exist beyond proof. Some computations never halt. Some infinities cannot be tamed. Even infinity itself had to be reimagined.Cantor showed that some infinities are bigger than others. **The infinity of real numbers is larger than the infinity of natural numbers.But the space between them— whether there's an intermediate infinity— remains undecidable.**It cannot be proven true. It cannot be proven false. It depends on the axioms you choose.

Logic, in its quest for certainty, found a deeper uncertainty.

Not ignorance— but limit. Not error— but boundary. We do not live in a world where all truths fit in a single system.We live in a world where systems must choose what to include and what to leave beyond their reach.**And in that choice, mathematics became not just a language— but a map of what can and cannot be said.**

Chapter 9

Computation – The Engine of Formal Thought

"To compute is to obey a logic so pure, it does not care who asks the question. It moves forward, step by step, blind to desire, deaf to doubt— yet from that silence emerges precision. An algorithm is not knowledge. It is the structure through which knowledge arrives."

9.1 The Algorithm as Essence

What is a thought if not a process?
What is understanding if not a sequence of steps from question to answer?

Computation begins when logic is set in motion.
When rules are not just stated, but followed.
At its heart, computation is a procedure. A path. An algorithm.

An algorithm is not a machine. It is the idea behind the machine.
A finite set of rules that—if followed— transform input into output.
It doesn't care who or what performs it. Human, machine, paper, bacteria—
Computation is substrate-agnostic.

Long before computers, algorithms shaped civilization.

- The ancient method to extract square roots
- The steps to convert fractions into decimals
- The Euclidean algorithm to find greatest common divisors

Each a finite recipe. Each a promise of repeatable outcome.

But in the 20th century, something changed.
We began to ask not just how to compute— but what could be computed.
Not just steps— but limits.

This required a new idea:
Formal procedures must be abstracted into a universal language.
Computation had to be defined with mathematical precision.
What counts as a step? What defines a halt? What can be decided, no matter how much time you give it?
The answers would redefine not just machines— but the nature of knowledge itself.

To ask what can be done, we first had to imagine a machine that could do anything.
And so, we arrive at the mind that mapped the edge of computability itself.
Alan Turing.

9.2 Turing's Universal Machine

In 1936, Alan Turing drew a map of what it means to follow instructions.

It was not a blueprint of any real device.

It was a theoretical engine— a symbolic creature defined only by rules.

A strip of infinite tape. A read/write head. A finite list of instructions.

And yet— this simple machine could simulate any process that could be formally described.

The Turing Machine was born.

Each square on the tape held a symbol— a mark, a blank, a bit.

The head scanned one square at a time, deciding what to do based on its current state and what it saw.

Write, move, change state— repeat.

From these modest parts, a storm of possibility arose.

The machine could count, sort, add, subtract, compare, simulate.

It could model chess.

It could compute π.

It could simulate itself.

But Turing didn't stop there.

He asked: **Could one machine simulate all other machines?**

The answer was yes.

A Universal Turing Machine is a machine that reads the description of any other machine and simulates it.

This idea was profound.

It meant that computation is not bound to specific tasks. **One structure can perform any process given the right code.** This is the foundation of modern computing. Your laptop. Your phone. Your brain simulation app. Your AI assistant. **All are, in essence, universal machines carrying out encoded procedures.** The machine is not what matters. **The program is. Turing separated hardware from logic. He**

separated matter from form. And in doing so, he revealed that computation is not a thing— **It is a relationship between steps and symbols.**

A dance of rules.**A thought, in motion.**

9.3 Halting, Oracle, Limit

The machine runs. It follows the rules. It processes the tape.
But what if it never stops?

Turing asked a question so simple, it sliced through computation like a blade:
Given a program and an input, can we always tell whether it will halt or run forever?
It seems reasonable. Machines are made to be understood.
But Turing proved otherwise.

There is no general method to determine whether a program halts.
This is the Halting Problem.

You cannot write a program that takes any other program and infallibly predicts whether it finishes.
Because if you could— you could feed it itself, invert its result, and create a contradiction.
The machine crashes on the question of its own behavior.

This limit is not a bug.
It is a feature of formal systems.
Even perfect logic meets boundaries that cannot be crossed.
There are problems that are well-posed, clearly stated— and yet undecidable.

Turing imagined another tool: **An oracle.**
A black box that could answer undecidable questions.
He didn't believe it could be built— only that it defined a new level of computation.

These are the degrees of uncomputability.
A staircase of problems rising beyond reach.
Some problems we can solve in time. Some only approximately. Some never at all.
Computation is not omnipotent.
It is powerful, but bounded.
And at its edge, we confront the shape of mystery not because of complexity— but because of structure.

9.4 Complexity: Not Everything Easy Is Fast

Just because something *can* be done doesn't mean it can be done *in time.*

Some problems are like weeds— solvable quickly.
Others are like forests— the path grows exponentially with every added tree.
Computation is not binary. It is not just *possible* or *impossible.* It is also *feasible* or *intractable.*

This is the realm of complexity theory.
It doesn't ask: **"Can it be done?"** It asks: **"How much does it cost?"**

We classify problems by the resources they demand:

- **P** — Solvable in polynomial time. Fast. Practical. Realistic.
- **NP** — Verifiable in polynomial time. But maybe not solvable quickly.
- **NP-complete** — The hardest of the hard. If you can solve one, you can solve them all.
- **EXP**, **PSPACE**, **EXPSPACE** — Towers of difficulty rising through time and memory.

And there, at the center of complexity theory, sits the unanswered riddle:
Is P = NP?
In simpler terms:

If you can check a solution quickly, can you always *find* one quickly?

It is a question that cuts across math, cryptography, biology, logic, and AI.
If **P = NP**, encryption collapses. Creativity could be automated. Proofs could be found in seconds.
If $P \neq NP$,
then effort still matters.
Then there are tasks we can only verify— but never shortcut.

Computation is shaped by cost.
A brute-force solution exists for many problems. But exponential time is a cage that no machine can break.
Not all solvable problems are solvable soon.
And so, even within the computable, limits persist.
Computation is not only a question of yes or no.
It is a question of when.

9.5 Computation Beyond Silicon

A computer is not a box. It is a process.
A pattern that unfolds when structure meets rules.

We built our first machines from metal, silicon, and wires.
But computation does not belong to circuits.
It belongs to form.

Today, computation escapes its original container.
It flows through new materials— new models— new metaphors.

Quantum computation uses qubits instead of bits—

not just 0 or 1, but superpositions.
It computes by evolving entire spaces of probability at once.
Its power lies not in speed— but in parallelism.

Biological computation uses cells, molecules, and genes as circuits.
Bacteria can compute logic gates. DNA strands can solve puzzles by folding into answers.

Neural computation models what brains do.
Not rule-by-rule logic, but weighted activations— learned, not programmed.
From this, machine learning emerges. Pattern recognition. Language. Art.

Unconventional computation pushes further:

- Soap films that solve minimization problems
- Slime molds that find shortest paths
- Spacetime geometries that compute as they evolve

Computation is a behavior. It can arise wherever structure can evolve predictably.

A tree's branching pattern. A planet's orbital stability. A fire's flickering chaos. Each obeys some rules. **Each computes something.** The universe itself— might it be a computation? **To compute is to transform. To transform is to express.** In this view, computation becomes not just a tool— but a fundamental mode of existence. And that means the question of what can be computed is also a question of what can be.

Chapter 10

Information – The Fabric of Communication

"Information is not a whisper from one place to another. It is the shape of contrast— a ripple across a field of possibilities, a mark that says: this is not that.
It does not travel as sound or symbol alone. It travels as difference that survives translation. And when meaning emerges, it is not because the signal was heard, but because it reduced uncertainty."

10.1 From Meaning to Measure

Before words, there were signals.
A glance. A drumbeat. A pattern of smoke.
Something happens here that matters there.
Communication is the transfer of difference.

But difference is not yet information. To formalize information, we needed to let go of *meaning*— and focus on structure. **Claude Shannon** did just that. In 1948, Shannon asked: How can we send a message through a noisy channel? He didn't care what the message meant— only how
uncertain it was. **The more unpredictable a message, the more information it carries.** If I tell you what you already know, I've communicated nothing. If I surprise you, I've reduced your uncertainty.
Information, in Shannon's theory, is not content. It is reduction in entropy. **Information is measured in bits.** Not "ones and zeroes"— but binary decisions. **How many yes-or-no questions would it take to guess the
message?** That's its information content. This shift turned meaning into mathematics. Words became probabilities. Messages became codes. Noise became quantifiable. Suddenly, we could optimize communication.
Compress language. Detect errors. Transcend noise.
Information became a physical quantity— not a human idea. And with it, a new way to see the world began: Not as substance, but as signal. Not as presence,

but as pattern.

10.2 Shannon's Entropy: Information as Surprise

We cannot measure meaning. But we can measure surprise. Imagine a coin flip. If the coin is fair, each outcome—heads or tails—has a 50% chance.
The uncertainty is high. You don't know what's coming. But what if the coin is rigged? Always lands on heads? Then there's no surprise. No information. **Certainty kills**
information. Claude Shannon captured this insight with a single, elegant formula:

$$H = -\sum_{x} p(x) \log_2 (p(x))$$

Where $p(x)$ is the probability of outcome x. the average amount of information produced by a source.
High entropy = high unpredictability = more information per symbol. Low entropy = predictability = less information.

Entropy is not chaos. It is surprise, properly measured. It shows up everywhere:

- In texts, where frequent letters (like "e") carry less information than rare ones (like "z").
- In genetics, where base pair patterns are measured by how much they deviate from randomness.
- In cryptography, where entropy protects secrets.

Shannon's entropy does not ask, "What does this mean?" It asks, "How hard is this to guess?" **The heart of communication is the management of uncertainty.** You do not just speak. You sculpt probabilities. **You encode, transmit, decode— all to shrink the unknown.** This realization— that information is the reduction of entropy— did not stay within telegraphs or radios. It began to spread— into thermodynamics, quantum mechanics, biology, economics, and neuroscience. Because wherever something can happen in multiple ways, information is hiding. Waiting to be extracted. Waiting to be compressed. Waiting to reshape the system that receives it.

10.3 Compression and the Limits of Knowledge

If you can say it with fewer bits, you understand it better.
Compression is not a trick. It is a mirror.
It reflects how much structure a message contains.
The more predictable the message, the more compressible it is.

Zip a file. You remove redundancy. You distill the essence. This is not just about saving space. It is about discovering pattern.

In 1965, Andrey Kolmogorov extended Shannon's insight.
He asked:
What is the shortest possible description of a string of data?
A truly random string has no shorter description than itself.
But a patterned string— say, "101010101010..."— can be expressed in a single rule.
Compression becomes a form of explanation.

This is the heart of Kolmogorov complexity:

$$K(x) = \text{length of the shortest program that outputs } x$$

The more compressible a pattern, the more structured it is.
And the more structure it holds, the more we can under-

stand.

But this power comes with a cost.
Kolmogorov complexity is uncomputable.
There's no general algorithm that can find the shortest description of an arbitrary string.
The limits of compression are the limits of provable insight.

Still, this framework reshapes how we view everything:

- A scientific law is a compression of data.
- A theory is a shortcut through experience.
- Intelligence is the ability to compress without loss.

Compression is not silence. It is precision.**It is the art of saying only what matters— and nothing more.**And so, information becomes not just a quantity, but a principle.**That which cannot be compressed is either noise— or truth.**

10.4 Signal, Noise, and Channel Capacity

No message travels alone. It rides through a world of distortion.

Noise is not error. It is variation— irrelevant change that threatens to overwrite the relevant.

Shannon's brilliance was not just in measuring information. It was in asking:

How much information can a channel carry, despite the noise?

Every communication path— a wire, a fiber, a brain— has a limit:

Its capacity.

Shannon defined it:

$$C = B \cdot \log_2(1 + \frac{S}{N})$$

Where:

- C is channel capacity,
- B is bandwidth,
- S is signal power,
- N is noise power.

This tells us: How much pure information can be sent reliably through a noisy medium.

If you exceed capacity, errors drown your meaning.

But if you stay beneath it, **you can communicate flawlessly— no matter how noisy the line.**

This gave rise to error-correcting codes.
They don't prevent noise. They anticipate it.
By spreading information cleverly, they allow receivers to reconstruct the message, even from damaged signals.

You use this every time you stream a video, send a text, or receive a space probe's whisper from millions of miles away.

Noise does not stop communication. It defines it.
The dance between clarity and chaos is not an obstacle— it is the medium.

To communicate well is not to remove noise, but to shape your signal so it survives it.

10.5 Information as Fundamental: Physics, Biology, Mind

At first, information was just a way to talk about phones, codes, and radios.
But something strange happened.
It kept showing up everywhere.

In **physics**, black holes don't just trap matter. They trap *information.*
Stephen Hawking's paradox was not about atoms— but about whether their information was lost forever.
The answer appears to be no.
The universe conserves information—
not just energy.

In **quantum mechanics**, particles are defined not by essence, but by the information they carry.
Entangled particles share information instantly.
The fabric of reality is structured as probabilities— waves of possible outcomes— collapsing into specifics only when observed.

In **biology**, life runs on code.
DNA is a chemical string, yes— but more deeply, it is a data structure.
It stores instructions, constraints, potential.

Genes are messages written in molecules.

In **neuroscience**, your thoughts, memories, perceptions— are not substance.
They are *patterns of information* propagating through neural networks.
You do not feel neurons. You feel the meaning of their configuration.

Across every level— matter, mind, machine— structure is secondary.
Information is primary. This idea— that the universe is not made of stuff, but of states— has become a radical new lens on reality itself. **"It from bit,"** as physicist John Wheeler said. **Reality is what can be said about what happens.** The world is not built from particles or forces alone— but from the relations between distinguishable alternatives. You are not just a body processing data. You are a structure through which information flows. And in that flow, form becomes thought. Noise becomes meaning. Difference becomes life.

Chapter 11

Networks – The Shape of Connection

"Nothing connects alone. A point becomes meaning only through its links. From neurons to nations, the pattern is the message."

The Node and the Edge – Anatomy of a Network

Before cities, before neurons, before the internet— there were connections.

A network is a structure of parts and the relationships between them. A node is a point. An edge is a connection. Together, they form a graph.

From family trees to traffic systems, from friendships to fungal root webs— graphs are the skeletons beneath the world's surface.

This is not metaphor. It is mathematics.

Graph theory, founded by Leonhard Euler in 1736, began with a puzzle: Could one walk across all seven bridges of Königsberg without retracing a step? The answer—no—did not lie in geography but in structure. Euler didn't need to walk the bridges. He needed only to abstract the system into nodes and edges.

That moment shifted the world.

It told us: The properties of a system can emerge entirely from how its parts are linked—regardless of what the parts are.

Every network has a topology—a pattern of connection. In a star, one node is the hub, radiating connections like a sun. In a ring, every node forms part of a loop, equal and continuous. In a mesh, every point overlaps redundantly, optimizing for fault-tolerance. In a tree, a single root branches outward, cascading dependencies.

Some networks are centralized. Others are distributed. Some amplify signals. Others buffer them.

The structure determines the flow.
Cut a hub in a star, and everything else goes dark.
Disrupt a ring, and only one link breaks. Fracture a tree near the base, and the entire canopy falls. But tear at a mesh, and the network reroutes, endures, remembers.
This is why nature favors redundancy.
The brain is not a single node, but a distributed mesh.
The internet survives war and disaster because its connections are many. Fungal networks beneath forests transfer nutrients from tree to tree— a biological internet older than language.
And in human society, networks define everything:
Who knows whom. Who influences whom. How ideas spread. How epidemics move.
Networks determine access, opportunity, and power. They reveal the hidden architecture of influence. The quiet geometry of belonging.
What matters is not merely how many connections exist— but where they are, how they form, how they cluster, how they transmit.
Social networks show homophily: birds of a feather cluster. Economic networks reveal bottlenecks: control is often held by few. Ecological networks exhibit resilience: loss of one species can be absorbed—until a threshold breaks.

Even within the brain, thought itself is a network phenomenon. Neurons fire in synchrony. Pathways strengthen through use. Ideas propagate not through a single flash, but through waves of connection across modular structures.

To understand anything complex— a market, a forest, a culture— is to understand the network beneath it.

And the most powerful truth: Networks evolve.

They grow. They rewire. They prune. They adapt. What starts as chaos finds form. What begins as chance solidifies into structure.And every structure leaves behind a pattern.

In networks, the pattern is not imposed from above. It emerges from the interactions of the parts.

Which means: You don't control a network by force. You shape it by changing how it connects.

In a connected world, the path matters as much as the node. The edge becomes the axis of possibility.

And so we arrive at a new kind of mathematics— one that doesn't just describe things, but maps their relationships.Not things, but links. Not positions, but flows.The more we understand networks, the more we understand the world—not as a set of isolated facts,but as a system of unfolding relationships. **Connection is destiny.** And the shape of connection shapes the future.

11.1 Centrality and Influence – Who Matters, and Why

Not all nodes are equal.
Some sit at the center. Some bridge between clusters.
Some connect many. Others only one.
In networks, influence is not declared— it is embedded.
Centrality is power by position.

There are many kinds:

- **Degree centrality** – How many direct connections a node has.
- **Closeness centrality** – How quickly a node can reach others.
- **Betweenness centrality** – How often a node sits on the shortest path between others.
- **Eigenvector centrality** – How connected a node is to other well-connected nodes.

Each reveals a different shape of importance.

A celebrity may have many connections (high degree). A diplomat may bridge groups (high betweenness). A wise elder may quietly sit in the shortest routes (high closeness). A trendsetter may influence influencers (high eigenvector).

Influence flows through structure, not just strength.

In social media, algorithms prioritize centrality. Who you see, who sees you— is shaped by invisible math.
In ecosystems, keystone species hold vast webs together. Remove them, and collapse ripples outward.
In biology, some neurons serve as critical hubs. Inactivate them, and cognition fails.

To shape a system, you do not need to control every part. You need to know where control concentrates naturally.
Not the loudest node. The most connected.
Not the biggest. The most between.
Not the highest. The most embedded.

Position is pattern. And pattern is power.

11.2 The Small-World Principle – Closeness in Vastness

The world feels big. But it's smaller than it seems.
In 1967, Stanley Milgram sent letters to random people in Nebraska. Their task: Get a letter to a Boston stockbroker— but only by passing it to someone they knew personally.
The result? On average, it took just six steps.
Six degrees of separation.

This is the small-world phenomenon.
Even in enormous networks, most nodes can be reached in surprisingly few steps.
Why?
Because random long-range links connect distant regions.

Mathematicians Duncan Watts and Steven Strogatz modeled this in 1998.

They showed that networks with a mix of local clustering and random shortcuts become small worlds.

High clustering, short path length.

Social networks. Neural networks. Power grids. Even proteins.

All display this structure.

It explains how:

- Diseases spread fast.
- Memes go viral.
- Ideas leap continents.
- Rumors race reality.

Connectivity is not just how many— but where.

One weak tie can collapse distance.

One forgotten bridge can unite two civilizations.

Scale does not equal remoteness.

In a small world, vastness is an illusion.

Everyone is closer than they appear.

11.3 Emergence and Phase Shifts — When Structure Wakes Up

A network is not just a map.
It is a field of potential— waiting for a threshold.
Then suddenly, everything changes.

This is emergence.
When local interactions give rise to global behavior.

A single neuron cannot think. But a brain can.
An ant is not intelligent. But a colony organizes itself.
A single computer is limited. But the internet became the largest distributed system ever built.

These are not metaphors. They are transitions.
At critical density, new phenomena appear.

In physics, this is called a phase shift.
Water becomes ice. Magnetism appears.
Synchronization erupts.
In networks:

- A few links—
  and the system is fragmented.
- Add more—
  a giant component forms.
- Add more—
  information flows.
- One more—
  the system becomes self-regulating.

Complexity is not added. It arises.
It is not in the parts. It is in their coordination.
And the coordination is not directed.
It is discovered.

The key is threshold.
Emergence is not a gradual slope. It is a cliff.
A tipping point. A flash of coherence.

This is why systems surprise us.
Nothing seems to change— until everything does.
Then suddenly, a new level of behavior appears.
Emergence is how networks wake up.
And once awake, they remember.

11.4 The Networked Self – Identity as Position

You are not just a mind in a body. You are a node in a system.
And your identity— your beliefs, memories, affiliations, behaviors— is shaped by your location in the network.

Sociologists call this **structural embeddedness**.
Your views are not isolated choices. They are signals received from those around you.

In the 1950s, Solomon Asch showed that people will conform to a group's opinion— even if it's obviously wrong.
But this wasn't weakness. It was structure.
In a network, truth becomes proximity-weighted consensus.

What you see, what you trust, what you fear— depends on your edges.

Algorithms amplify this.
Social media filters don't just reflect preferences. They reinforce them.
You are shown what fits your neighborhood.
Echo chambers are not beliefs.
They are architectures.

Even your sense of "self" is network-sensitive.
Neuroscience reveals that the brain doesn't house a single narrator.
It is a coordination center, where subnetworks stabilize into a coherent story.
Your personality is the attractor state of inner networks.

This applies to culture too.
Languages, customs, values— spread not by logic, but by connection.
To exist is to be positioned.
Not above, not outside— but within.

You are not the source of your influence. You are its conduit.
And through you, the network shapes its next expression.

Chapter 12

Category – The Mathematics of Meaning

To understand a thing, we must look not at what it is,
but how it relates.
Not at its shape, but its transformations.
Not at the object, but the arrows it sends and receives.

In category theory, meaning does not reside within.
It emerges between.

12.1 Beyond Objects – Arrows as First-Class Citizens

In most of mathematics, we start with objects.
A number. A point. A set. An equation.
But what if we started somewhere else?
Not with the things themselves— but with the relationships between them?

Category theory begins not by asking what something is, but what it does.
It is a mathematics of behavior.

A **category** is defined by:

- A collection of **objects**
- A collection of **arrows** (also called morphisms) between them
- Each arrow has a source and a target
- Arrows can be composed (if the target of one matches the source of another)
- Composition is associative
- Every object has an identity arrow

That's it.

You can build a universe from this.
The objects don't matter— only how they connect.

Think of sets and functions: Sets are objects, functions are arrows.
Think of spaces and continuous maps. Think of groups and group homomorphisms. Think of logical systems and proofs.

Category theory does not eliminate detail— it abstracts it.
And in doing so, it reveals when two wildly different systems are behaving in the same structural way.
It replaces essence with equivalence.

This is not shallowness. It is depth redefined.
When you stop staring at objects, you start to see what they mean to each other.
Category theory isn't about ignoring complexity. It's about rising above it— until only the relationships remain.
Where algebra sees a function, category sees an arrow.
Where topology sees a space, category sees what maps preserve. This is why it's often called "mathematics of mathematics"— because it tells us how different branches are reflections of one another.
In categorical thinking, two structures are "the same" if they behave identically from the outside.
We call this an **isomorphism**— not sameness,
but structural compatibility.
You don't ask "What is this?" You ask "What does this do?"
Meaning is function.

This perspective is powerful.

It allows computer scientists to write software that obeys algebra.

It allows physicists to model systems with identical behavior despite different ingredients.

It allows logicians to map proofs between systems that speak different symbolic languages. **Every arrow is a message about consistency.**In a world overwhelmed by detail,category theory simplifies without reducing.It gives us tools to recognize the same patterns in places that look unrelated.This is not erasure. It's integration.**By focusing on arrows,you don't lose the**

objects. You understand them better Because a thing is known not in isolation— but in its connections. This is the gift of category theory: **It shows that mathematics is not a pile of facts—but a web of relationships.**

And those relationships can be traced, composed, understood.

In arrows, we find a different kind of truth— not in what something is, but in what it becomes.

12.2 Functors – Mapping Structures Between Worlds

Once you have a category— a web of objects and arrows— you can compare it to another.
But how?
Enter the functor.
A **functor** is a map from one category to another that preserves structure.
It sends:

- Each object in the source category to an object in the target
- Each arrow (or morphism) to an arrow in the target
- Identity arrows to identity arrows
- Composition of arrows to composition of images

Think of a functor as a structure-respecting translator. It doesn't just move things. It keeps their relationships intact.

For example:

- A functor from the category of groups to the category of sets (forgets structure, but keeps the underlying set)
- A functor from topological spaces to groups (creates algebraic invariants, like the fundamental group)
- A functor from logic to computation (recasting proofs as programs)

Functors allow entire systems of meaning to be mapped, translated, compared.
They are not just bridges— they are mirrors.
Through functors, we learn that two categories are speaking the same language— in different dialects.

And when functors form patterns of their own— we build categories of categories.
Meta-structure arises.
Mathematics starts folding inward— reflecting on itself.

12.3 Natural Transformations – Harmony Across Layers

If categories are worlds, and functors are bridges— what happens when we compare bridges?

We arrive at the next layer: **natural transformations.** **A natural transformation** is a way of relating two functors that go from the same source category to the same target category.
It assigns to each object a morphism (arrow) between the outputs of the two functors— in a way that respects the structure of the category.

Said more poetically: if functors are translations between languages, natural transformations are ways the translations can smoothly deform into each other without breaking meaning.

For each object A, if F and G are functors, a natural transformation η gives a morphism:

$$\eta_A : F(A) \to G(A)$$

such that for every morphism $f : A \to B$ in the original category, the following diagram commutes:

$$F(A) \xrightarrow{\eta_A} G(A)$$
$$\downarrow F(f) \qquad \downarrow G(f)$$
$$F(B) \xrightarrow{\eta_B} G(B)$$

In plain terms: no matter how you go— transform then map, or map then transform— you end up in the same place.

That's naturality. It is a kind of higher-level symmetry.

Where functions tell you how elements behave, functors tell you how structures behave, natural transformations tell you how behaviors relate.

This layering is not complexity for its own sake. It's precision.

In a world made of systems, we must understand not just what connects— but how those connections themselves can align.

The higher the layer, the deeper the harmony.

12.4 Universal Properties – Essence Without Content

What makes something what it is?
Not its material. Not its name. Not its coordinates.
But how it relates to everything else.

This is the insight behind **universal properties**.
In category theory, an object can be defined not by what it is inside— but by how everything else points to or from it.
Essence becomes role.

For example: the product of two objects A and B is not just a pair (A, B). It is an object P together with two projection morphisms:

$$\pi_1 : P \to A \quad \text{and} \quad \pi_2 : P \to B$$

such that for any other object X with morphisms $f : X \to A$ and $g : X \to B$, there exists a unique morphism $u : X \to P$ that factors both f and g through P:

$$\pi_1 \circ u = f \quad \text{and} \quad \pi_2 \circ u = g$$

This triangle of relationships **uniquely characterizes the product**.
No reference to internal structure. Only external coherence.

This pattern repeats:

- Products (as above)

- Coproducts (dual to products)
- Initial and terminal objects
- Limits and colimits
- Pullbacks and pushouts

Each is defined by a universal property— a minimal and unique relation to the rest of the system.

The object is determined by the arrows around it.
This is radical.
It means structure can arise without specification.
It can be *characterized* by its necessity in a pattern.
Not what it is— but what must point to it for everything else to work.

This is mathematics as architecture.
And the blueprints are made of arrows.

12.5 Category as Philosophy – Mathematics Reflecting on Itself

At the surface, category theory seems like a formal tool— a language for organizing complex relationships.
But deeper down, it becomes something else:
Philosophy in mathematical form.

Most of mathematics asks about content.

What is a number? What is a shape? What is a space?
Category theory asks about form.
Not "what is this?" but "what role does it play?"
Not "what is inside?" but "how does it connect?"

It turns mathematics into a study of behavior, context, and transformation.
It shifts focus from objects to interactions.
This is not reduction. It is elevation.

In this view:

- Sets are not containers—
  they are participants in functions.

- Numbers are not quantities—
  they are points in algebraic flow.

- Proofs are not just chains—
  they are morphisms in logical space.

- Theorems are not isolated facts—
  they are invariants under transformation.

Mathematics becomes a study of harmony.
Of how systems preserve structure while changing form.
Of how meaning persists under translation.

Category theory, then, is more than a branch of math.
It is mathematics as reflection.
A way for math to describe itself.

It is not about the furniture inside the universe— but the grammar of how anything can be a universe at all.

Chapter 13

Topology – The Mathematics of Shape Without Distance

"What matters is not how far, but how connected."

A coffee cup and a donut are the same to a topologist.

Because where geometry sees size, topology sees continuity.

It asks not how things measure— but how they hold together when stretched, bent, or curled.

This is not about distance. It is about essence.

13.1 Topological Spaces – When Stretching Is Allowed

Most people think of geometry as the math of shapes and distances.
But what if distance didn't matter?
What if you could stretch a shape like rubber— bend it, twist it— as long as you never tear it?

This is topology.
And its basic unit is not a point, but a sense of nearness.
A topological space is a set of points, along with a collection of "open sets" that satisfy three rules:

- The whole space and the empty set are open
- Any union of open sets is open
- Any finite intersection of open sets is open

That's it.
But from this minimal structure, emerges a universe of connectivity.

Open sets are the essence of locality. They tell you what neighborhoods look like. They define
continuity, closeness, and structure— without relying on measurement.

In topology, two spaces are "the same" if they can be continuously transformed into each other.
No cuts. No glue. No tears.
Just smooth deformations— like melting clay into a new shape.

So a triangle is the same as a circle. A square is the same as an ellipse. A coffee cup is the same as a donut.
Topology doesn't ask, "How big is it?" It asks, "How is it connected?"

Continuity becomes the guiding principle. A function between spaces is continuous if the inverse image of every open set is open.
This simple condition preserves the fabric of connection between domains.

Homeomorphism is the gold standard of sameness. Two spaces are homeomorphic if there exists a continuous bijection between them with a continuous inverse.
This is not about precise coordinates— it is about what survives transformation.

Holes matter. The number of holes in a space cannot be smoothed away.
A sphere is not a torus. A torus is not a figure eight. These are topologically distinct.

And so we count holes.

The Euler characteristic tells us something fundamental about a shape's structure:

$$\chi = V - E + F$$

Vertices minus edges plus faces. This invariant holds across all polyhedral deformations.

Topology lets us classify spaces not by their appearance, but by their underlying form.
It becomes the mathematics of essence— what cannot be changed.
When measurement falls away, connection remains.

Even when the world twists, even when the boundaries blur, some truths persist.
Topology captures these.
Not the length of a road, but whether it loops. Not the width of a river, but whether it divides. Not the shape of a surface, but whether it holds.
It teaches us that form is not fixed— but relationship is.
That you can stretch, compress, mold— but some structures refuse to break.
To see through deformation, and glimpse what lasts.

13.2 Homeomorphism – Identity Through Deformation

What does it mean for two shapes to be "the same"?
In Euclidean geometry, it means congruence— same angles, same lengths.
But topology is softer. More forgiving. More structural.
Two spaces are topologically the same if one can be stretched, bent, or twisted into the other— without cutting or gluing.
This is called a **homeomorphism.**

Formally, a homeomorphism is a function:

$$f : X \to Y$$

such that:

- f is bijective (one-to-one and onto)
- f is continuous
- f^{-1} is also continuous

If such a function exists, X and Y are **homeomorphic**. They may look different— but topologically, they are the same.

A circle is homeomorphic to a square. A sphere is homeomorphic to an ellipsoid. A donut (torus) is homeomorphic to a rubber tube joined at the ends.

But a sphere is *not* homeomorphic to a donut.

Why?

The hole.

Topology cares deeply about holes.

Not their shape. Not their size. Just their existence.

In this view, identity becomes a kind of transformational symmetry.

Not what something looks like, but what it can become— without breaking.

The world is full of objects that differ on the surface, but share the same deeper form.

Topology teaches us to see through the distortion.

To look for the invariants beneath the bending.

13.3 Open Sets and Neighborhoods

—

The Breath of Space

In topology, space is not measured— it is **felt**.
We don't ask how far one point is from another. We ask: **What surrounds this point?** What kinds of open sets wrap around it like a gentle atmosphere?

This is the topology's core tool: **open sets**.
An open set is a kind of "breath" in space— a zone that contains a point and everything sufficiently close to it, but never its boundary.
You are never on the edge of an open set. You are immersed in it.

A **topological space** is defined by the collection of all such open sets, which satisfy three simple axioms:

- The whole space and the empty set are open.
- Any union of open sets is open.
- Any finite intersection of open sets is open.

These rules may seem abstract, but they unlock powerful concepts:

Continuity: A function is continuous if the preimage of every open set is open.

Convergence: A sequence converges if its tail fits within every open neighborhood of the target point.

Interior and Boundary: The interior of a set consists of all points with some open set inside it. The boundary is what's left— where closeness is contested.

Topology gives us tools that transcend coordinates. We don't ask how large or small— we ask what surrounds, what persists, what cannot be removed without rupture. This changes everything.

We can describe the shape of a molecule without measuring every bond. We can study the shape of the universe without a ruler long enough to span it. We can detect holes, voids, and loops by how space is connected— not how it's measured.

The open set becomes the voice of the unseen.
Not edges, but neighborhoods. Not exactness, but inclusion. Not distance, but continuity.

Topology becomes a language for resilience. A way to speak when measurement fails. A whisper through which space breathes its own selfhood.
This breath of space lets us model everything from neural networks to black holes.
Why?
Because everything— from atoms to abstractions— exists in relation to what surrounds it.
Topology teaches us to think in terms of containment, not just location. In terms of possibility, not just position.
Where analysis ends, topology begins.
It does not care how you stretch the paper, so long as it does not tear. It does not care about numbers, only neighborhoods.
In the absence of rulers, we still find rules.
In the dissolution of measure, we discover form.
And so, the open set is not merely a mathematical object— it is how space speaks to itself.

13.4 Connectedness, Compactness, and Continuity

Once you've given space the power to breathe with open sets, you can ask deeper questions:
Is this space in one piece? Can it be covered with small parts? Do functions behave smoothly across it?
Topology answers these with three deep concepts: **connectedness**, **compactness**, and **continuity**.

Connectedness asks: Can the space be split into two disjoint open sets?
If yes, it's disconnected. If no, it's **connected**— a single whole.
The real number line is connected. Two separate circles are not. The space in your coffee mug remains connected even as it swirls.

Compactness is subtler. It means: Every open cover has a finite subcover.
Intuitively, you can't spread infinitely thin and still blanket the space.

- The closed interval $[0, 1]$ is compact.
- The open interval $(0, 1)$ is not.

Compactness gives you control over infinity.
It ensures limits exist, minima are attained, and functions behave nicely.

Continuity reemerges— but now topologically defined.
A function $f : X \to Y$ is continuous if the preimage of every open set in Y is open in X.
This echoes the intuition of "no breaks, no jumps."
But it doesn't require derivatives, nor coordinates.
Just open sets.

These three concepts— how space stays whole (connectedness), how it stays bounded (compactness), and how functions flow across it (continuity)— form the pillars of topological structure.
They let us reason about convergence, transformation, and identity— in any context where shape matters more than size.
Topology is where form survives the stretching.

13.5 Applications – From Knots to the Cosmos

Topology may seem abstract— but it pulses underneath the world.
Its principles shape not just pure mathematics, but physics, biology, computer science, and beyond.

Knots and Braids: In topology, a knot is not a tangle of rope— it is a closed loop in 3D space, studied up to continuous deformation.
Knot theory helps biologists understand how DNA coils and uncoils. It helps chemists study molecular chirality. It even impacts quantum computing.

The Shape of the Universe: General relativity tells us that spacetime is curved. But topology asks: What is its global shape?
Is it finite or infinite? Does it loop back on itself? Are there topological defects in the fabric of space?
Cosmologists use topology to model possible geometries of the entire universe.

Topology in Data Science: Topology can find shape in data without needing coordinates.
Persistent homology, a tool from **topological data analysis**, detects clusters, loops, and holes in high-dimensional datasets— revealing structure that statistics alone
cannot.

Network Theory and Neuroscience: Topological invariants help map brain networks, analyze social systems, and quantify resilience in systems under change.

Phase Transitions and Fields: In physics, topological properties remain invariant under smooth deformations— leading to new kinds of matter like topological insulators.
Even the behavior of particles can be constrained by the topology of their environment.

Computation and Logic: Topology underlies many models of computation and proof. It is the silent grammar beneath continuity and constructibility.

Topology is everywhere— not because it measures, but because it listens.
It hears what remains when shape bends but essence stays.

Chapter 14

Algebra – The Language of Symmetry

"Structure is not found in what things are,
but in how they act.
Algebra does not study objects—
it studies agreements,
the silent contracts between operations
that keep a system coherent."

14.1 Algebraic Structures – Operations That Behave

Algebra begins not with numbers, but with rules.
What matters is not the objects themselves, but how they interact.
What happens when you combine them? Can you undo that combination? Is the outcome predictable?
These are questions of **structure**.

An **algebraic structure** is a set equipped with one or more operations that satisfy certain laws.
Here are the most basic:

- **Group**: one operation (like addition or multiplication)
  with an identity, associativity, and inverses.
- **Ring**: two operations (like addition and multiplication),
  with one forming a group and the other distributing over it.
- **Field**: a ring where every nonzero element
  also has a multiplicative inverse.

These structures are not defined by the elements— they are defined by the **rules those elements obey**. Whether you are working with numbers, rotations, matrices, or polynomials— what matters is whether the structure holds.

These rules give rise to astonishing power.
They let you:
- Prove that some equations can never be solved
- Understand the symmetry of crystals and molecules - Encrypt secrets through algebraic cycles - Build error-correcting codes for deep space communication

Algebra is the art of describing what remains when things change.
It is the mathematics of balance, reversibility, and harmony.
It's not about solving equations— it's about the systems those equations live in.

14.2 Groups – The Anatomy of Symmetry

A group is one of the most fundamental ideas in mathematics— and one of the most powerful.
At first glance, it's just a set of elements with a way to combine them.
But if that combination follows four simple rules, something extraordinary happens:
Structure emerges. Patterns stabilize. Symmetries unfold.

Formally, a **group** is a set G with a binary operation $\cdot$ satisfying:

- **Closure**: For all $a, b \in G$, the result $a \cdot b \in G$.
- **Associativity**: For all $a, b, c \in G$, we have $(a \cdot b) \cdot c = a \cdot (b \cdot c)$.
- **Identity**: There exists $e \in G$ such that $a \cdot e = e \cdot a = a$ for all $a \in G$.
- **Inverse**: For each $a \in G$, there exists $a^{-1} \in G$ such that $a \cdot a^{-1} = a^{-1} \cdot a = e$.

That's it.
But from this minimal setup arises the foundation of symmetry.
Every time you rotate a shape, permute a set, invert a matrix, shift a waveform— you are witnessing a group in action.

Examples of groups:
- The integers under addition - The nonzero real numbers under multiplication - Rotations of a regular polygon - Permutations of elements in a list
Each captures a different kind of symmetry— but all obey the same rules.

Groups are not just about numbers. They model:
- The algebra of transformations in geometry - The structure of particles in quantum physics - The modular arithmetic behind cryptography
Group theory is the mathematics of reversible structure.
It tells you what can be transformed without being destroyed.
It is the hidden grammar of motion, repetition, and design.

14.3 Rings and Fields – Number Systems Reimagined

A group is powerful— but it has only one operation. When you add a second operation, and define how they interact, you enter the world of **rings** and **fields**.

A **ring** is a set equipped with two binary operations, typically called addition and multiplication, satisfying:

- Addition forms an abelian (commutative) group.
- Multiplication is associative.
- Multiplication distributes over addition:

$$a \cdot (b + c) = a \cdot b + a \cdot c$$

A ring is like the integers: you can add, subtract, and multiply— but not always divide.

A **field** goes further.
It is a ring in which:

- Every nonzero element has a multiplicative inverse.
- Multiplication is commutative.

In a field, you can do all four arithmetic operations without restriction—except division by zero.
The real numbers, rational numbers, and complex numbers are fields.
So is every finite field— used in cryptography, error correction, and computation. Why does this matter? Because once you define two operations, you can build: -

Polynomials - Matrices - Modular systems - Vector spaces - Algebraic equations

And more importantly— you can start asking: Which properties remain when equations are transformed? This leads to the towering power of algebraic structures in solving ancient problems: - Proving impossibility of angle trisection - Showing that general fifth-degree equations cannot be solved by radicals - Defining prime ideals and algebraic varieties - Powering the mathematics behind elliptic curve cryptography Rings let us model complexity. Fields let us solve within constraints.

Both expand the world from mere numbers to structured systems.

A ring is a universe of operations. A field is a playground of reversibility.

Together, they let us explore how systems behave when layered with operations that both combine and constrain.

In rings, not everything is invertible— and that's where the richness begins.

In fields, everything flows until the singularity of zero stops it.

Algebra is no longer arithmetic. It is structure with laws.

And those laws shape the tools that shape our world.

From polynomial equations to digital signals, from error-correcting codes to the hidden structure of primes—

Rings and fields are how we extend the reach of symmetry into the heart of systems.

14.4 Homomorphisms – Maps That Preserve Meaning

What if you could take one algebraic system and map it onto another— without losing its structure?
That is the power of a **homomorphism**.
A homomorphism is not just a function. It is a function that respects the rules.
It says: "If I know how elements behave over here, I can predict how they'll behave over there."

Formally:

- A **group homomorphism** $f : G \to H$ satisfies:

$$f(a \cdot b) = f(a) \cdot f(b)$$

- A **ring homomorphism** preserves both addition and multiplication:

$$f(a + b) = f(a) + f(b), \quad f(a \cdot b) = f(a) \cdot f(b)$$

What does this mean in practice?
It means structure can travel.
You can compress, reframe, re-encode while keeping the essence.

A homomorphism is a promise that behavior is preserved across transformations.

These maps are the language of mathematical equivalence.
They help define:
- Isomorphisms: perfect structural matches - Kernels: what gets "flattened" by the map - Quotients: new structures formed by collapsing sameness

In group theory, homomorphisms classify symmetry. In field theory, they define algebraic extensions. In number theory, they help build Galois theory. In computer science, they underlie data transformations and logic
systems.

Homomorphisms show that even as systems shift, some truths remain stable.
Structure is not place-bound. It is portable. Meaning can be mapped.

14.5 The Reach of Algebra – From Crystals to Codes

Algebra may begin in symbols— but it does not stay there.
Its structures reach into the deepest workings of nature, technology, and logic.

In physics: The symmetries of space and time are described by groups.
Noether's theorem connects every symmetry to a conservation law.
Lie groups and Lie algebras model the continuous symmetries underlying particle physics.

In chemistry: Crystals are classified by their symmetry groups.
Molecules obey algebraic constraints that determine chirality, reaction paths, and structural stability.

In computer science: Algebra powers:

- Finite fields in encryption (e.g., AES, RSA)
- Polynomial rings in error-correcting codes
- Group theory in public key cryptography
- Boolean algebra in logic circuits

In communication: Reed–Solomon codes, used in CDs, QR codes, and deep-space transmissions, rely on finite field arithmetic to detect and correct errors with algebraic precision.

In pure math: Algebra is the bridge between number theory, geometry, and logic.
Galois theory uses field extensions to prove that the quintic has no general formula.
Category theory reframes all mathematics in terms of objects and morphisms— generalized homomorphisms.

In AI and data science: Tensor algebra structures neural networks. Linear algebra powers learning algorithms. Abstract algebra informs pattern matching, automata, and semantics.

Algebra is not just one subject. It is the **grammar of structure** underlying many disciplines.
It teaches us:
- How systems behave under transformation - What properties are preserved - How meaning travels between forms

Algebra is not just about solving for x. It is about understanding what x belongs to— and how it behaves when the world shifts.
It is the mathematics of elegance under constraint, and power under abstraction.

Chapter 15

Linear Algebra

When numbers learned to point, they became
vectors.
And with direction came dimension—
not just how much,
but where,
how fast,
and toward what future.

15.1 Vectors – Quantities with Direction and Magnitude

A number tells you how much. A vector tells you where and how much.
It is not just a value— it is a direction in space.

In one dimension, numbers suffice. But the moment you need to move in a plane, navigate a volume, or model forces and flows—
you need vectors.

A vector is an ordered list of numbers:

$$\vec{v} = \begin{bmatrix} v_1 \\ v_2 \\ \vdots \\ v_n \end{bmatrix}$$

Each component represents a coordinate in a particular direction.
Together, they describe position, force, velocity, or any quantity with magnitude and orientation.

Vectors can be:

- Added: $\vec{u} + \vec{v}$ — tip-to-tail.
- Scaled: $c \cdot \vec{v}$ — stretching or shrinking.
- Subtracted: $\vec{v} - \vec{u}$ — displacement.
- Projected: onto a direction or subspace.

They obey simple rules:
- Commutativity: $\vec{u}+\vec{v} = \vec{v}+\vec{u}$ - Associativity: $(\vec{u}+\vec{v})+\vec{w} = \vec{u}+(\vec{v}+\vec{w})$ - Distributivity: $c(\vec{u}+\vec{v}) = c\vec{u}+c\vec{v}$

These rules define a **vector space**— a set of vectors with operations of addition and scalar multiplication that follow those laws.

Linear algebra studies these spaces and the transformations between them.
It's not just about solving equations— it's about understanding:
- How spaces are shaped - How they change - And what stays the same under transformation
A vector tells you where. Linear algebra tells you what happens when you go there.

15.2 Matrices – Machines That Move Space

If vectors describe where you are going, matrices describe how space itself moves.

A matrix is not just a table of numbers. It is a transformation.

It scales, rotates, reflects, shears, or compresses— and it does so with ruthless precision.

Formally, a matrix is a rectangular array:

$$A = \begin{bmatrix} a_{11} & a_{12} & \cdots & a_{1n} \\ a_{21} & a_{22} & \cdots & a_{2n} \\ \vdots & \vdots & \ddots & \vdots \\ a_{m1} & a_{m2} & \cdots & a_{mn} \end{bmatrix}$$

It acts on a vector $\vec{v}$ by multiplication:

$$A\vec{v} = \vec{w}$$

The result $\vec{w}$ is another vector, transformed by the "machine" A.

Matrix operations mirror physical actions:

- Multiply by a rotation matrix → rotate space
- Multiply by a scaling matrix → expand or shrink space
- Multiply by a shear matrix → skew space

Matrices can be composed—like chaining machines together. They can be inverted—reversing the transformation. They can be transposed—flipping dimensions.
They are the operators of linear systems.

In practical terms, matrices are used to:
- Solve systems of linear equations - Transform images and 3D models - Rotate objects in computer graphics - Encode information in cryptographic systems - Optimize processes in data science and machine learning

Every matrix encodes a rule: how inputs become outputs, how shapes become shadows, how causes lead to effects. **In the world of vectors, matrices are the engines. They are how form is changed— and structure is revealed.**

15.3 Linear Systems – When Equations Become Geometry

A single equation is a line. Two equations intersect—perhaps. Three may define a point. Or they may contradict, or align completely.
What we once saw as rows of symbols, linear algebra sees as intersecting planes in space.

Consider a system of linear equations:

$$2x + y = 5$$
$$3x - 2y = 4$$

This is not just a problem to solve— it is a geometric configuration.
Each equation is a line in the plane. Their solution, if it exists, is their intersection point.

More generally:

$$A\vec{x} = \vec{b}$$

Where:
- A is a matrix of coefficients - $\vec{x}$ is a vector of unknowns
- $\vec{b}$ is the target output
The goal: find the vector $\vec{x}$ that, when transformed by A, becomes $\vec{b}$. **This is linear algebra's power: solving algebraic systems by understanding geometric space.** Three possibilities arise:

- **One solution:** the planes intersect at a unique point
- **Infinite solutions:** the planes coincide
- **No solution:** the planes are parallel and never meet

These outcomes reflect the **rank** of the matrix, the dimension of the space spanned by its rows or columns. Gaussian elimination, matrix inversion, row-reduction— these are not just techniques. They are **moves within a space**, searching for alignment, contradiction, or degrees of freedom. Linear systems appear everywhere:
- Circuits with resistors and voltages - Economic models of supply and demand - Resource allocation in logistics - Solving networks of interdependent variables What once seemed like algebra becomes geometry in motion.
Solving a system is not about finding x and y. It's about understanding the shape of a possibility space— and locating yourself within it.

15.4 Eigenvectors and Eigenvalues – The Invariant Directions

Most vectors change when transformed by a matrix. They rotate, stretch, or compress.
But some vectors resist.
They stay on their own line— only scaled, not redirected.
These are the **eigenvectors**.
And their scale factors are the **eigenvalues**.

Formally, for a square matrix A:

$$A\vec{v} = \lambda\vec{v}$$

Where:
- $\vec{v}$ is the eigenvector - λ is the eigenvalue
This equation says: "The transformation A turns $\vec{v}$ into a version of itself, scaled by λ."

Not all vectors are eigenvectors. Only those that lie along privileged directions— invariant under the transformation.
But every matrix that acts on a space has its own secret symmetry: a set of directions that do not shift, even as everything else moves.

Why do these matter?
Because eigenvectors and eigenvalues:

- Reveal the natural modes of a system
- Identify stable directions in dynamics

- Compress complex transformations into simple components
- Power principal component analysis in data science
- Govern vibration modes in physics and engineering

In a vibrating string, eigenvectors define the fundamental tones. In Google's PageRank, the dominant eigenvector ranks importance.
In machine learning, they isolate the axes of variance.

To find them, we solve:

$$\det(A - \lambda I) = 0$$

This **characteristic equation** produces the eigenvalues. Each leads to its own eigenvector.
They are the DNA of the transformation— its essence, expressed geometrically.
They tell us what stays the same, even as everything else changes.

15.5 The Power of Linearity – Why It Works Everywhere

Why does linear algebra appear in nearly every scientific field, every engineering discipline, every data analysis pipeline?
Because linearity is the closest thing to simplicity the universe allows.

A linear relationship is one where:
- Inputs combine predictably - Scaling inputs scales outputs - The parts determine the whole
It obeys:

$$f(a\vec{u} + b\vec{v}) = af(\vec{u}) + bf(\vec{v})$$

This may look humble— but it allows systems to be modeled, predicted, transformed, and optimized.

Linearity is the threshold between chaos and order.
Go nonlinear, and all bets are off— but stay linear, and you can solve, simulate, and stabilize.

That's why:

- Physics begins with linear approximations
- Circuits rely on linear components (Ohm's Law)
- Machine learning uses linear transformations between layers
- Signal processing filters frequencies using linear systems
- Quantum mechanics is built on linear operators and vector spaces

Even when systems are nonlinear, we linearize them—zooming in to a flat patch of an otherwise curved world.
Linear algebra gives us:
- Efficient representations of data (via matrices)
- Understanding of dimensionality and rank - Projection onto meaningful subspaces - A toolkit for rotations, transformations, decompositions
It teaches us to:
- See structure in dimension
- Simplify complexity
- Transform without losing what matters

It is not just a field. It is a worldview.
The belief that, beneath the apparent complexity, there exists a form— expressible by lines and combinations—that we can understand.
Linear algebra is the compass that keeps direction even when the map changes.

Chapter 16

Statistics – Patterns in the Noise

In a world of shadows and samples,
statistics does not promise truth—
only clarity under uncertainty.
It listens through the noise,
measures what flickers,
and teaches us to act before
knowing everything.

16.1 What Is Statistics? From Description to Decision

You never get the full picture.
You get fragments, samples, snapshots.
Still, you must decide.

Statistics is the art and science of inferring the unseen from the seen. It does not demand certainty. It thrives in uncertainty. From the earliest weather forecasts to today's models of pandemics and markets, statistics offers a framework for belief under partial information.

A sample is not the world, but it whispers about it.
You compute a mean— and catch a glimpse of a center. You measure variation— and feel the spread. You track correlations— and uncover the gravity that binds variables together.

These are not certainties. They are signals.
And with them, you decide.

Statistics moves in two stages:

- **Descriptive:** What is the shape of the data?
- **Inferential:** What might be true beyond what we see?

A histogram compresses a thousand numbers into one sweeping curve. A standard deviation gives shape to unpredictability. A confidence interval offers a window not on what *is*, but on what *could be* with plausible likelihood.

But the real power lies not in summaries— but in decisions.

Should we act now? Is this result significant? Is this difference real, or just random? Statistics cannot answer these perfectly. But it can help us ask better questions. Every model is a simplification.

Every test carries assumptions. Yet in the dance between data and doubt, statistics gives us balance. We build hypotheses, knowing we might reject the truth or accept a fluke.

We model trends, knowing they may bend.

We update beliefs— Bayes whispering in the background— each new data point a nudge toward refinement.

To do statistics is to live honestly with the partial. It is to accept that clarity does not require completeness.It is to act, while still learning.In science, it checks claims.In medicine, it weighs treatments.

In justice,

it detects bias.In business, it forecasts risk.And in daily life, it reminds us:**Evidence matters. Assumptions shape outcomes. And probability is not weakness— it is wisdom.**In the end,statistics is not about being right. It is about being less wrong,more often,with humility and with method.

16.2 Descriptive Statistics – What the Data Shows

Before you analyze, predict, or infer— you describe.
You look at what's there, and try to capture its essential form.
Descriptive statistics tells you:
- Where the data is centered - How much it varies - What its shape is - Whether patterns or outliers exist

Central Tendency — What is "typical"?

- **Mean** (average): The sum of all values, divided by count
$$\mu = \frac{1}{n}\sum_{i=1}^{n} x_i$$
- **Median**: The middle value when sorted
- **Mode**: The most frequent value

Each captures "center" differently. In symmetric distributions, these measures may align. In skewed distributions, they diverge— the mean is pulled toward the tail, while the median stands firm, and the mode clings to the peak.
Understanding the center helps frame what's "normal"— but only if you also know how data deviates from it.

Dispersion — How much do values spread?

- **Range**: Maximum minus minimum

- **Variance**: Average squared deviation from the mean

$$\sigma^2 = \frac{1}{n}\sum_{i=1}^{n}(x_i - \mu)^2$$

- **Standard deviation**: The square root of variance

- **Interquartile range (IQR)**: The middle 50% of the data

Dispersion is not noise. It is the breath of the data—how far its lungs expand when it inhales the world.
If center tells you where most values live, dispersion tells you how far they roam.

Shape — How does the distribution behave?
- Is it symmetric or skewed? - Does it have heavy tails or a sharp peak? - Is it unimodal, bimodal, or something stranger?
The shape of data determines the rules of engagement. A bell curve obeys different expectations than a power law. A long tail holds surprises that averages obscure.

Outliers — What doesn't fit?
Descriptive tools help reveal them: Z-scores, IQR cutoffs, and graphical clues. But an outlier isn't always an error. Sometimes it is a signal. A mutation. A revolution waiting to be noticed.

Correlation — Do variables move together?

$$r = \frac{\mathrm{Cov}(X, Y)}{\sigma_X \sigma_Y}$$

A correlation of $r = 1$ means perfect alignment. $r = 0$ means independence. But even strong correlation is not a chain of cause— just a mirror of motion.
Look closely. Two trends dancing in sync might be strangers in essence.

Descriptive statistics is not about prediction. It is the art of reflection.
It helps you listen before you speak.
It makes no assumptions. It forms no theories.
But it gives the shape on which all theories must stand.
They are the anatomy of data— the form beneath the function.
To describe well is to begin knowing.

16.3 Inference – Learning from Samples

You almost never see the whole.
You see a slice— a sample— a fragment of a world you must understand.
Statistical inference asks: If this is what we observed, what can we say about what we didn't?

Let $X_1, X_2, \ldots, X_n$ be a random sample from a larger population.
We want to use these n values to estimate:
- The population mean μ - The true proportion p - The spread, shape, or other hidden parameters
But we don't just want to estimate— we want to estimate **with confidence**.

Confidence Intervals provide a range:

$$\bar{X} \pm z^* \cdot \frac{\sigma}{\sqrt{n}}$$

Where:
- $\bar{X}$ is the sample mean - σ is the population standard deviation (or sample std. dev. s) - z^* is a multiplier from the standard normal distribution - n is the sample size
This interval says: "There is a 95% chance the true mean lies within this range."

Hypothesis Testing compares competing claims:
Suppose you suspect a coin is biased. You observe 8 heads out of 10 flips.
Is that unusual enough to reject the idea of fairness?
Steps:

1. Set the **null hypothesis** (H_0: the coin is fair)
2. Choose an **alternative hypothesis** (H_1: the coin is biased)
3. Compute a test statistic (e.g., z or t)
4. Calculate the p-**value**: probability of seeing data this extreme if H_0 is true
5. Decide: Reject H_0 if $p < \alpha$ (e.g., 0.05)

p-**value** is not the probability that H_0 is true. It is the probability, under H_0, of getting what you saw—or worse.

Type I error: Rejecting a true H_0 **Type II error:** Failing to reject a false H_0
Inference navigates the unknown by quantifying doubt.
It is not about certainty— but disciplined uncertainty.

16.4 Regression – Modeling Relationships

Some variables don't live alone.
They move together, pull on each other, signal each other's tendencies.
Regression is how we capture that co-movement— how we turn a cloud of points into a predictive model.

The simplest case is **linear regression**.
We want to model a relationship like:

$$Y = \beta_0 + \beta_1 X + \epsilon$$

Where:
- Y is the dependent variable - X is the independent variable - β_0 is the intercept - β_1 is the slope - ϵ is the error term (the leftover noise)

Using data $(x_1, y_1), (x_2, y_2), \ldots, (x_n, y_n)$, we find the line that minimizes the total squared error:

$$\text{Minimize} \sum_{i=1}^{n} (y_i - \hat{y}_i)^2$$

Where $\hat{y}_i = \beta_0 + \beta_1 x_i$ is the model's prediction.
This gives us the **least squares solution**— the best-fitting line.

Interpretation:
– β_1 tells us how much Y changes for each unit of X
– The **coefficient of determination**, R^2, tells us how much of Y's variation is explained by X

$$R^2 = 1 - \frac{\mathrm{SS}_{\mathrm{res}}}{\mathrm{SS}_{\mathrm{tot}}}$$

Residuals (errors between actual and predicted values) help assess fit and diagnose problems. Patterns in residuals signal that something is missing— perhaps the relationship isn't linear, or key variables have been left out. **Multiple regression** extends the model:

$$Y = \beta_0 + \beta_1 X_1 + \beta_2 X_2 + \cdots + \beta_p X_p + \epsilon$$

Now, we fit a plane (or hyperplane) in higher dimensions. Regression becomes powerful not just as a line, but as a lens.It gives you a way to say: "If I know this, I can estimate that." "If this changes, here's what likely follows." **Regression turns observed patterns into usable structure— a formula from the fog.**

16.5 Bayesian Thinking – Learning as Updating Beliefs

What do you believe?
Not in the philosophical sense— but in the statistical one. Bayesian thinking begins with a prior belief and updates it based on new evidence.

At the core is **Bayes' Theorem**:

$$P(H \mid D) = \frac{P(D \mid H) \cdot P(H)}{P(D)}$$

Where:
- H is a hypothesis - D is observed data - $P(H)$ is the prior probability - $P(D \mid H)$ is the likelihood - $P(H \mid D)$ is the posterior probability - $P(D)$ is the total probability of the data

This formula answers:
Given what I believed before, and what I just saw, what should I believe now?
Prior + Evidence → Posterior

This process is recursive.
After each update, the posterior becomes the new prior.
Belief flows forward through time.

Example: Medical Diagnosis
Suppose a disease affects 1 in 1000 people.
You take a test that's 99% accurate.
You test positive.
What's the chance you actually have the disease?

Most guess: close to 99%.
But using Bayes' Theorem:

- $P(\text{Disease}) = 0.001$
- $P(\text{Positive} \mid \text{Disease}) = 0.99$
- $P(\text{Positive} \mid \text{No Disease}) = 0.01$

Then:

$$P(\text{Disease} \mid \text{Positive}) \approx \frac{0.99 \cdot 0.001}{0.99 \cdot 0.001 + 0.01 \cdot 0.999} \approx 0.09$$

Only 9%! Why? Because the disease is rare— and false positives outnumber true ones. Bayesian thinking resists overreaction. It tempers confidence. It balances the old and the new. **It treats belief not as binary, but as a continuous update.**
It makes reasoning probabilistic. In machine learning, Bayesian models estimate parameters as distributions, not point values.
In science, they model uncertainty in the face of sparse data.
In daily life, they whisper: "Hold your belief loosely— until the next observation." Every observation is a nudge. Not a command, but a shift. It pulls belief gently—toward better alignment.
Bayesian thinking is not about being right. It is about being less wrong. It is the mathematics of humility.
Where classical logic draws hard lines between true and false, Bayesian logic introduces a spectrum: Truth becomes a distribution. Certainty becomes a curve.

You do not leap to conclusions. You migrate—statistically. What you knew becomes weighed against what the world just revealed. Together, they form what you now hold likely.

This is how a child learns language. How a scientist adjusts a theory. How a driver anticipates traffic. How the brain builds its model of the world.

Bayesian inference is not just a formula. It is the scaffolding of adaptation. It mirrors evolution itself: A prior gene pool. A selective environment. A posterior population.

The difference is speed. Biology adapts across generations.

Bayesian models adapt with every signal.

The result? A belief system in motion. A structure that reshapes as reality unfolds. Not behind or ahead of the world—but beside it.

And the most powerful part: It never throws anything away. Old beliefs are not deleted. They are softened. Weighted. Held in dynamic tension with the new.

$$\text{Belief}_{t+1} = \text{Belief}_t \cdot \text{Evidence}_t$$

This is not prediction. It is alignment under uncertainty.

So the next time your confidence swells, ask: Have I updated? Or am I clinging to a prior that the world has already left behind?

Chapter 17

Combinatorics – The Art of Counting

Counting is not just tallying.
It is seeing structure in possibility,
order in chaos,
and constraint in freedom.
Combinatorics is the mathematics of choice—
of paths not taken,
and of every way they could be.

17.1 What Is Combinatorics? From Counting to Configuration

How many ways can you arrange a deck of cards?
How many passwords are possible with six characters?
How likely is it that two people in a room share a birthday?

Combinatorics is the mathematics of counting— but not mere counting.
It's not about how much. It's about how many ways.
It explores the structure of possibility, the patterns of arrangement, and the logic of discrete configuration.

At its core, combinatorics deals with:

- **Permutations** – Ordered arrangements
- **Combinations** – Unordered selections
- **Partitions** – Ways to break a whole into parts
- **Recurrence relations** – Counting defined recursively
- **Inclusion-exclusion** – Correcting for overlaps
- **Pigeonhole principle** – Structure in inevitability

These tools let us answer questions like:
- How many different routes exist between cities? - In how many ways can a set be formed? - What's the probability of a specific hand in poker?

But combinatorics is more than clever formulas.
It is the foundation of:
- Cryptography - Algorithm design - Graph theory - Coding theory - Statistical mechanics
And it begins with the simplest of questions: **"In how many ways?"**
Counting becomes a lens— revealing hidden order in even the most chaotic-seeming arrangements.

Each arrangement is not just a possibility. It is a configuration of structure. A fingerprint of constraint and freedom intertwined.
You can shuffle a deck of cards in 52! ways— an unimaginable number: $52! \approx 8.07 \times 10^{67}$
That's more than the atoms in the Earth. More than the seconds since the Big Bang. Each shuffle is a universe—plausible, precise, and profoundly unique.

But combinatorics is not only about the vast. It thrives in the small.
Three coins. Two dice. A handful of colors.
From these, it builds a cathedral of logic. Where possibility branches. And symmetry hides in plain sight.

In permutations, order matters. A lock code is not the same backwards. In combinations, order fades. A salad does not care how you added the lettuce.

Partitions divide the whole. In how many ways can a number be broken into smaller ones? This is not just arithmetic— it is the language of balance, of composition, of structure.

Recurrence relations whisper: Today's count depends on

yesterday's structure. And the day before that. Like Fibonacci—where the past feeds the future.

Inclusion-exclusion sharpens precision: Add all the parts. Subtract the overlaps. Then add back the ones double-subtracted.
Mathematics, like nature, respects nothing more than accuracy.

And the pigeonhole principle—deceptively simple—declares: Put $n + 1$ items in n boxes, and at least one box must overflow. From this, entire theorems unfold. Combinatorics reveals a strange truth: Constraint is not a limitation. It is the birthplace of pattern. The structure of all potential lies not in chaos— but in how you arrange what already exists. Where algebra gives rules, combinatorics gives possibilities. Where calculus studies flow, combinatorics studies form. Each count is a map of what could be. Each formula is a telescope into structure. So when you ask, "How many ways?" you are not just asking about numbers. You are asking: What shapes does logic allow?

17.2 Permutations and Factorials

—

When Order Matters

Suppose you have 3 books. How many ways can you place them on a shelf?
Try it:
- ABC - ACB - BAC - BCA - CAB - CBA
That's 6 arrangements.
Why?
Because for the first position, you have 3 choices. Then 2 choices remain. Then 1.

$$\text{Total permutations} = 3 \cdot 2 \cdot 1 = 3! = 6$$

Factorials generalize this:

$$n! = n \cdot (n-1) \cdot (n-2) \cdots 1$$

Examples:
- $4! = 4 \cdot 3 \cdot 2 \cdot 1 = 24$ - $5! = 120$ - $0! = 1$ (by definition, for consistency in formulas)

Permutations without repetition:
If you choose k elements out of n and arrange them, the count is:

$$P(n,k) = \frac{n!}{(n-k)!}$$

Example: How many 3-letter passwords can you make from 5 distinct letters (no repeats)?

$$P(5,3) = \frac{5!}{(5-3)!} = \frac{120}{2} = 60$$

Permutations with repetition:
If some items are repeated, the count changes.
Example: How many ways to rearrange the word "LEVEL"?
- 5 letters - L appears twice, E twice, V once

$$\frac{5!}{2! \cdot 2! \cdot 1!} = \frac{120}{4} = 30$$

Permutations model:
- Seating arrangements - Race rankings - Passwords and codes - Anagrams
Where order matters, permutations rule. And factorials are their currency.

17.3 Combinations – When Order Doesn't Matter

If you pick 3 students to form a team, does it matter who you picked first?
No.
Combinations count selections where order is irrelevant.

Notation:

$$\binom{n}{k} = \frac{n!}{k!(n-k)!}$$

This is called a **binomial coefficient**. It reads: "n

choose k".

Example: How many ways to choose 3 students from a class of 10?

$$\binom{10}{3} = \frac{10!}{3! \cdot 7!} = \frac{720}{6} = 120$$

Combinations appear in:

- Lottery odds - Committees - Poker hands - Groupings - Subset counts

Every time order doesn't matter, combinations answer the question.

Key idea:

Every group of 3 students (ABC) has 6 permutations (ABC, ACB, BAC, BCA, CAB, CBA) but counts as just **one combination**.

That's why we divide by $k!$ in the formula— to eliminate overcounting.

Pascal's Triangle elegantly encodes combinations:

$$\binom{n}{k} = \binom{n-1}{k} + \binom{n-1}{k-1}$$

Each entry is the sum of the two above it.

Combinations also appear in algebra:

$$(a+b)^n = \sum_{k=0}^{n} \binom{n}{k} a^{n-k} b^k$$

This is the **binomial theorem**— and it links counting to algebraic expansion.

Combinations are the structure beneath choice— where identity matters, but order does not.

17.4 Inclusion-Exclusion – Correcting for Overlaps

Suppose you're counting how many students play **soccer**, **basketball**, or **chess**.
Some students play more than one.
How do you avoid counting them twice?
Inclusion-Exclusion fixes overcounting by subtracting overlaps.

For two sets:

$$|A \cup B| = |A| + |B| - |A \cap B|$$

Three sets:

$$|A \cup B \cup C| = |A| + |B| + |C| - |A \cap B| - |A \cap C| - |B \cap C| + |A \cap B \cap C|$$

General formula:

$$|\bigcup_{i=1}^{n} A_i| = \sum |A_i| - \sum |A_i \cap A_j| + \sum |A_i \cap A_j \cap A_k| - \cdots + (-1)^{n+1} |A_1 \cap A_2 \cap \cdots \cap A_n|$$

Example:

Out of 100 students:

. 60 like pizza (P) . 50 like tacos (T) . 30 like both $(P \cap T)$

How many like at least one?

$$|P \cup T| = |P| + |T| - |P \cap T| = 60 + 50 - 30 = 80$$

Why it matters:

This technique shows up in:

- Counting unique users with overlapping accounts - Survey analysis - Set operations in databases - Probability of union events

Inclusion-exclusion is a precise scalpel in the chaos of overlapping categories.

It prevents overcounts, finds exclusives, and clears up the muddy water of intersecting sets.

17.5 Pigeonhole Principle – Guarantees from Scarcity

If you place 13 socks into 12 drawers, at least one drawer must contain two socks.
Obvious? Yes. Profound? Also yes.

The Pigeonhole Principle: If you place more items than containers, at least one container must hold more than one item.

Formally: If n items are placed into m containers and $n > m$, then at least one container holds at least $\lceil n/m \rceil$ items.

Example:
In a room of 367 people, at least two share a birthday. Why? There are only 366 possible birthdays (including leap day). So with 367 people, overlap is certain.

Applications:
- In number theory, to prove existence of divisibility - In data hashing, to guarantee collisions - In computer science, to prove limits of compression - In geometry, to prove proximity among points

Extended version: If 10 people each pick a number from 1 to 3, some number must be picked at least 4 times. Because:

$$\left\lceil \frac{10}{3} \right\rceil = 4$$

The power of the principle is not in what it predicts— but in what it forbids.

It forbids the idea that scarcity can avoid repetition. It forbids perfect separation when space runs out. It forbids uniqueness in overloaded systems.

It proves the inevitable, by showing that even randomness can't escape structure.

The Pigeonhole Principle does not predict which drawer overflows. It only says: *one must.*
That is its quiet power. It offers no location. No identity. Only inevitability.

This is not probability. This is logic without exception. It is the scaffolding beneath surprise. A law so firm, even chaos must obey.

It tells you:

- In any group of 27 words, at least two must begin with the same letter—because there are only 26.
- In any sequence of 100 integers from 1 to 50, at least two must repeat.
- In any function from a larger finite set to a smaller one, collisions are guaranteed.

In compression, this becomes an upper bound: Not every file can shrink. There are only so many shorter outputs. Too many inputs means—some must share code.

In passwords, in proofs, in packing— the Pigeonhole Principle stands behind impossibility. It says: You cannot fit an ocean into a glass without spilling some over.

Its real lesson? Some things cannot be avoided. Not because of bad planning, but because of mathematics.

It teaches: If you want separation, you must make room. If you want uniqueness, you must leave space. If you want precision, you must embrace limits.

Even randomness bows to this. No amount of shuffling can escape the arithmetic of overflow.

The Pigeonhole Principle does not seek control. It reveals constraint. It does not point to order. It points to repetition.

And in that repetition, a new structure begins— not from intention, but from certainty.

Chapter 18

Set Theory – The Foundation of Mathematics

Before numbers, before logic,
there was the question: Who belongs?
Set theory begins not with counting,
but with gathering—
with drawing boundaries around meaning.
It is the mathematics of inclusion,
of belonging,
of the invisible line between "in" and "out."

18.1 What Is a Set? Defining by Belonging

A set is not a thing. It is a rule of inclusion.
A set is a collection defined by what is inside it— and what is not.

Examples:
- The set of prime numbers less than 10: $\{2, 3, 5, 7\}$ - The set of vowels: $\{a, e, i, o, u\}$
- The set of days in a week: $\{\text{Mon}, \text{Tue}, \ldots, \text{Sun}\}$
If something is in the set, we write: $x \in A$ If it is not: $x \notin A$

Sets are everywhere.
They define:
- Number systems - Functions - Spaces - Events in probability - Structures in logic - Nodes in networks

Types of sets:
- **Finite set:** $\{1, 2, 3\}$ - **Infinite set:** $\{1, 2, 3, \ldots\}$ - **Empty set:** $\emptyset$ — a set with nothing in it - **Singleton:** a set with exactly one element, like $\{0\}$ - **Universal set:** everything under consideration

Notation:
- Union: $A \cup B$ — all elements in A or B - Intersection: $A \cap B$ — elements in both - Complement: A^c — everything not in A - Difference: $A - B$ — in A, not in B - Subset: $A \subseteq B$ — all members of A are in B

To define is to delimit. To understand is to compare. But to delimit is also to decide— to draw a line between what matters and what doesn't.
This is not just mathematics. It is judgment, cloaked in precision.
Every set is a choice— an act of distinction.
To build a theory, you must first decide which elements belong.
Does zero count? Is the boundary inclusive? Can a set contain itself?
These are not trivial. They led to paradoxes. They cracked foundations.
Russell asked: What about the set of all sets that do not contain themselves?
That question broke the naive view. Set theory had to rebuild itself— with axioms, with care.

The Zermelo-Fraenkel axioms (now with the Axiom of Choice) give us a stable ground.
From them we build numbers, functions, spaces, and logic.
They are scaffolding for all of modern math. They tell us what sets can exist, and how they relate.
But deeper still— set theory is not just the foundation.
It is the gesture of separation.
It is how we say: This is inside. That is not.
And in that gesture, we create meaning.
We sort the infinite into manageable ideas.
Set theory is structure beneath all structure.
It is the unseen sorting that lets order emerge.
To gather, to divide, to label, to link— this is the beginning of mathematics.
And perhaps, of thought itself.
Set theory gives us the tools to:
- Classify - Relate - Separate - Combine - Structure
It is the grammar of mathematics— a language of membership.
It all begins with a question: What counts as part of this?

18.2 Operations on Sets – Union, Intersection, and Beyond

Set theory is not just about what belongs— but what overlaps, separates, or emerges when sets are combined.

1. Union – $A \cup B$
All elements in A, or in B, or in both.

$$A = \{1, 2, 3\}, \quad B = \{3, 4, 5\} \Rightarrow A \cup B = \{1, 2, 3, 4, 5\}$$

2. Intersection – $A \cap B$
Elements common to both sets.

$$A \cap B = \{3\}$$

3. Set Difference – $A - B$
Elements in A but not in B.

$$A - B = \{1, 2\}$$

4. Symmetric Difference – $A \Delta B$
Elements in either A or B, but not both.

$$A \Delta B = (A - B) \cup (B - A) = \{1, 2, 4, 5\}$$

5. Complement – A^c
Everything not in A (relative to some universal set U)

$$U = \{1, 2, 3, 4, 5\}, \quad A = \{1, 2\} \Rightarrow A^c = \{3, 4, 5\}$$

Set operations obey laws:
- **Commutative:** $A \cup B = B \cup A$, $A \cap B = B \cap A$ - **Associative:** $(A \cup B) \cup C = A \cup (B \cup C)$ - **Distributive:** $A \cap (B \cup C) = (A \cap B) \cup (A \cap C)$ - **De Morgan's Laws:**

$$(A \cup B)^c = A^c \cap B^c, \quad (A \cap B)^c = A^c \cup B^c$$

These operations are not mere formalities. They form the basis of:
- Boolean logic - Probability events - Database queries - Logic circuits - Computer science foundations

To manipulate sets is to manipulate meaning itself.

18.3 Subsets and Power Sets – The Structure of Possibilities

To understand a set, you must know what smaller sets lie within it.

Subset: $A \subseteq B$ means every element of A is also in B. If A is strictly smaller: $A \subset B$ (proper subset)

The empty set $\emptyset$ is a subset of every set. Every set is a subset of itself.
Example:

$$S = \{a, b\} \Rightarrow \text{subsets: } \emptyset,\ \{a\},\ \{b\},\ \{a, b\}$$

Power Set: The set of all subsets of a set S. Denoted $\mathcal{P}(S)$.

$$\text{If } |S| = n, \quad |\mathcal{P}(S)| = 2^n$$

Example: If $S = \{x, y, z\}$, then:

$$\mathcal{P}(S) = \{\emptyset,\ \{x\},\ \{y\},\ \{z\},\ \{x, y\},\ \{x, z\},\ \{y, z\},\ \{x, y, z\}\}$$

Why Power Sets Matter:
- In logic: every proposition has truth assignments (subsets of possibilities) - In computing: permissions, states, and configurations use power sets - In probability: the sample space is often a power set - In category theory and topology: power sets form foundational structures

Each set holds a shadow realm— an architecture of all possible substructures.
The power set is the ghost of possibility haunting every defined collection.
It is not just what is, but what could be.
Every subset is a version of the whole— a shadow cast by removing parts.
This is not subtraction. It is variation. It is contingency made formal.

In the power set, you see the combinatorial explosion that hides beneath simplicity.
Three elements? Eight subsets. Ten elements? Over a thousand.
With every added point, possibility doubles. The power set scales not linearly, but exponentially. It is the mathematics of choice and of imagination. Each subset asks: What if only these remained? And the total collection asks a deeper question: What if every path were preserved? In a power set, we do not merely count. We listen to the chorus of combinations.
It is the full fingerprint of a finite universe.
It tells us not just what is present— but how presence itself can be arranged.
To understand a set is to trace its power set— the branching structure of its inner logic.
A kingdom of partial truths.
A map of potential order.
A lattice beneath the known.

And in that lattice, the architecture of mathematics takes root.

18.4 Infinite Sets – Beyond Counting

Some sets never stop. And yet, we can compare their sizes.

Finite vs. Infinite:
- $\{1, 2, 3\}$ – finite - $\mathbb{N} = \{1, 2, 3, \dots\}$ – infinite
But not all infinities are equal.
Countable infinity: If the elements of a set can be matched 1-to-1 with the natural numbers, it is countably infinite.

$$\mathbb{N}, \ \mathbb{Z}, \ \mathbb{Q} \quad \text{are countable}$$

Uncountable infinity: Some sets are so large that even listing them forever is not enough.

$$\mathbb{R} = \text{all real numbers}$$

Between 0 and 1, there are uncountably many real numbers. You cannot list them all, even in infinite time.**Cantor's Diagonal Argument:** Shows that any attempt to list all real numbers will always miss some— because new numbers can be constructed by diagonally altering decimal digits. **Cardinality of infinite sets:** - Countably infinite: $\aleph_0$ (Aleph-null) - Uncountable (real numbers): has greater cardinality

$$|\mathbb{N}| = \aleph_0 \quad < \quad |\mathbb{R}|$$

Why this matters:

- In calculus and real analysis - In understanding the continuum - In logic and philosophy of mathematics - In computer science (limits of computability) **Infinity is not one thing— but a layered hierarchy. Set theory reveals how even the immeasurable can be compared.** It dares to sort the unsortable. To give structure to what has no bound. Infinity is not chaos. It has tiers, and between each tier— a chasm no count can cross. From $\aleph_0$ to the continuum, and beyond— each level of infinity is an entire scale of reality. There are sets larger than the real numbers. There are infinities stacked upon infinities— **uncountable towers reaching into abstraction.** But even here, mathematics does not collapse. It refines. Cantor showed us that you can climb an infinite staircase without ever reaching a final step. The idea of size transforms. Not how much, but how listable. Not count, but correspondence. Two sets are the same size if their elements can be matched, even if neither has an end. This is the subtle art of comparing the endless. It is not measurement. It is mapping. **Infinity does not defy math— it defines it.**And once you allow the infinite into your system, you begin to see:**Some limits are not boundaries— they are invitations.**

18.5 Foundations and Paradoxes — Where Definitions Collapse

Set theory was built to clarify mathematics. But its own rules led to contradictions.

Russell's Paradox: Let R be the set of all sets that do not contain themselves. Does R contain itself?
- If it does, then by definition it should not. - If it doesn't, then by definition it should.
A logical contradiction.
This paradox shattered naive set theory.

The Response: Axiomatic Set Theory
To avoid paradoxes, mathematicians formalized rules of set construction.
Zermelo-Fraenkel Set Theory (ZF): A rigorous system with rules like:
- **Extensionality:** Sets are defined by their members. - **Foundation:** No infinite descent of sets containing sets.
- **Separation:** You can only form subsets from existing sets. - **Replacement:** New sets arise only through definable functions.
With the Axiom of Choice (AC), this becomes ZFC—the most widely used foundation of mathematics today.

Gödel's Incompleteness Theorems:
Even with axioms,
there are limits.
Gödel showed that:

- Any system rich enough to describe arithmetic must contain true statements it cannot prove.
- Consistency cannot be proven from within the system.

What this means:
- Mathematics has boundaries - Logic has blind spots - No foundation is complete
But paradox is not failure. It is a guide.

The power of set theory is not that it solved everything— but that it exposed the edge between clarity and chaos.
In that edge, mathematics finds its humility.

Chapter 19

Mathematical Logic

Every proof begins with a quiet wager:
that truth can be built,
one step at a time.
That from silence and symbols,
clarity can emerge—
not by magic,
but by rule.
Logic is not opinion.
It is structure laid bare.
It is the grammar of certainty
in a world full of noise.

19.1 Propositions and Truth Values – The Atoms of Logic

Mathematics rests on proof, and proof begins with logic. But logic itself begins with one act: **asserting that something is either true or false**.

A proposition is a declarative statement that is either true or false, but not both.
Examples:
- "2 is an even number" — True - "All primes are divisible by 2" — False - "If it rains, then the ground is wet" — a conditional proposition
Truth value:
- True $= T = 1$ - False $= F = 0$

Basic logical connectives:

- **Negation** $(\neg P)$: not P
- **Conjunction** $(P \wedge Q)$: P and Q
- **Disjunction** $(P \vee Q)$: P or Q
- **Implication** $(P \to Q)$: if P, then Q
- **Biconditional** $(P \leftrightarrow Q)$: P if and only if Q

Each of these can be described using a **truth table**— a method of mechanically computing every possible case.

Truth Table for Implication ($P \to Q$):

P	Q	$P \to Q$
T	T	T
T	F	F
F	T	T
F	F	T

Note: An implication is only false when a true premise leads to a false conclusion.

This is the atomic level of reasoning.
With it, we build proof. With it, we formalize argument. With it, mathematics becomes a language that thinks.

19.2 Logical Equivalence and Proof – Transforming Truth

Not all expressions look the same. But some mean the same.

Logical equivalence captures this idea:

$$P \equiv Q \quad \text{means that } P \leftrightarrow Q \text{ is always true}$$

Two propositions are logically equivalent if their truth tables are identical.

Examples:

$$\neg(P \wedge Q) \equiv (\neg P \vee \neg Q)$$

(De Morgan's Law)

$$P \to Q \equiv \neg P \vee Q$$

(Implication as disjunction)

$$\neg(P \to Q) \equiv P \wedge \neg Q$$

Proof techniques:

Logic provides structure for building proofs.

- **Direct Proof:** From P, deduce Q
- **Contrapositive:** Prove $\neg Q \to \neg P$
- **Proof by Contradiction:** Assume the opposite and show contradiction

- **Disjunctive Syllogism:** If $P \vee Q$ and $\neg P$, then Q
- **Modus Ponens:** If $P \to Q$ and P, then Q
- **Modus Tollens:** If $P \to Q$ and $\neg Q$, then $\neg P$

Formal proof:
A sequence of steps, each justified by a logical rule, leading from assumptions to conclusion.

Example:

Premise 1: $P \to Q$Premise 2: $Q \to R$Conclusion: $P \to R$

This is **hypothetical syllogism**, a valid chain of reasoning.
Logic is not about truth alone— it is about transformation.
It reveals the paths by which truth travels, and the scaffolding that holds it in place.

Logic is not about truth alone— it is about transformation.
It reveals the paths by which truth travels, and the scaffolding that holds it in place.

But even scaffolding can disguise itself.
A statement may wear many masks— some complex, some elegant, yet beneath the surface, their structures align.
This is the power of equivalence: to strip away decoration, to expose the underlying form.

A proof is not a performance. It is a procession of consequence. Each step must follow—not suggest, not imply— but follow.

When we reason through logic, we are not persuading. We are revealing inevitability.
Every rule, every inference, is a move within a closed system— a dance choreographed by consistency.
And once a conclusion is reached, it cannot be un-reached unless an assumption was flawed.
Logic does not care how you feel. But it will show you, if you listen, what must be.

19.3 Predicate Logic – Describing the World with Variables

Propositional logic is a start— but it cannot say:
- "Every number has a successor." - "Some birds cannot fly." - "No prime number is even—except 2." To describe these, we need logic with variables. **Predicate logic** expands the language of truth. **Predicates:**
A predicate is a statement template— a function that becomes a proposition when variables are filled. Example:

$$P(x) : \text{"x is a prime number"}$$

$$P(2) = \text{True}, \quad P(4) = \text{False}$$

Quantifiers:

- **Universal quantifier:** $\forall x$ — "for all x"
- **Existential quantifier:** $\exists x$ — "there exists x"

Examples:
- $\forall x \in \mathbb{N},\ x+1 > x$ - $\exists x \in \mathbb{Z},\ x^2 = 4$ These quantify over a domain of discourse— they let logic talk about entire collections. **Combining logic and quantifiers:**

$$\forall x (P(x) \to Q(x))$$

Means: If $P(x)$ is true for a given x, then $Q(x)$ must also be true for that x.

Negating quantified statements:

$$\neg(\forall x\ P(x)) \equiv \exists x\ \neg P(x)$$

(It's not true for all $\Leftrightarrow$ It's false for some)

$$\neg(\exists x\ P(x)) \equiv \forall x\ \neg P(x)$$

Predicate logic powers:
- Foundation of formal mathematics - Description of computer programs - Modeling statements in databases - Basis of formal proof assistants - The syntax of automated reasoning
Where propositional logic sees truth, predicate logic sees structure.
It lets mathematics speak about anything with clarity and constraint.

19.4 Incompleteness and Decidability – The Limits of Logic

Logic was built to eliminate doubt. To anchor mathematics on irrefutable ground.
But what if the anchor slips?

Gödel's Incompleteness Theorems (1931) shook the foundations.
In any consistent formal system rich enough to describe arithmetic:

- There are true statements that cannot be proven within the system.
- The system cannot prove its own consistency.

This wasn't a flaw in logic. It was a feature of reality.

Proof Sketch:
Gödel encoded statements as numbers— "arithmetized syntax." He created a proposition that said, "This statement is not provable."
If it were provable, it would be false. If not provable, then it's true—but unprovable.

Turing and Undecidability:
In 1936, Alan Turing showed there is no general algorithm to decide whether any given program halts. This is the **Halting Problem**— and it proves that not all questions can be answered mechanically.

Other undecidable problems:
- Does a Diophantine equation have integer solutions? - Is a given logical formula universally valid? - Can a computer program produce a particular output?

The result:
- There is no complete formal system for mathematics. - There is no perfect machine for decision. - Even logic must bow to limits.

But from these limits rises a new kind of certainty:
That mathematics is not static. It is alive— not just in what it knows, but in how it discovers what cannot be known.

19.5 Logic and Computation

The truths of logic are not just for mathematicians—they are for machines.

Propositions become programs. Proof becomes execution.

From logic to code:

- Propositional logic $\to$ Boolean circuits
- Predicate logic $\to$ Rule-based systems
- Proof $\to$ Computation

The Curry–Howard Correspondence: A profound insight:

$$\text{Proof} \quad \leftrightarrow \quad \text{Program}\text{Proposition} \quad \leftrightarrow \quad \text{Type}$$

Under this view, writing a proof is like writing a program that transforms assumptions into conclusions.

Logic programming: Languages like Prolog are built entirely on predicate logic.

You define relationships and facts. The machine deduces what follows.

Example:

"‘‘prolog parent(alice, bob). parent(bob, carol). ancestor(X, Y) :- parent(X, Y). ancestor(X, Y) :- parent(X, Z), ancestor(Z, Y). The machine can answer: "Who is an ancestor of Carol?"

Formal verification: Logic checks the correctness of code. • Are there any bugs? • Are all edge cases covered? • Can this program be proven to terminate?

Mathematical logic powers tools that verify the security, safety, and behavior of real-world systems.

Proof assistants: Coq, Lean, Isabelle — systems that help humans write formal, machine-checkable proofs.

From Aristotle to Algorithms, logic has become computation.

It no longer just describes thought— it performs it.

Chapter 20

Category Theory – The Mathematics of Mathematics

Mathematics asks what a thing is.
Category theory asks how it behaves.
It does not name the substance—
it traces the structure.
Not the object, but the arrow.
Not the essence, but the echo of connection.

20.1 What Is a Category? Objects, Arrows, and Composition

In most of mathematics, we study things: numbers, shapes, sets, spaces.
In category theory, we study how things relate.

A **category** is a structure made of:

- **Objects** — abstract entities (they could be sets, spaces, types...)
- **Morphisms** (or arrows) — directed relationships between objects

Every morphism has a direction:

$$f : A \to B$$

Read: "f is a morphism from A to B."

Two rules must always hold:

1. **Composition:** If $f : A \to B$ and $g : B \to C$, then there is a composite morphism $g \circ f : A \to C$
2. **Identity:** For every object A, there is an identity morphism $\mathrm{id}_A : A \to A$, such that:

$$\mathrm{id}_B \circ f = f = f \circ \mathrm{id}_A$$

These two rules capture an enormous range of mathematical behavior.

Examples of categories:

- **Set** — objects are sets, morphisms are functions
- **Grp** — objects are groups, morphisms are group homomorphisms
- **Top** — objects are topological spaces, morphisms are continuous maps
- **Cat** — objects are categories themselves, morphisms are functors

What matters in category theory is not the nature of the objects— but the web of arrows that connect them. **Mathematics, seen categorically, is no longer a pile of truths. It is a map of transformations.**

20.2 Functors – Maps Between Categories

In set theory, we map elements. In category theory, we map entire structures.
A **functor** is a morphism between categories. It preserves the shape of how objects connect.

Definition: A functor $F : \mathcal{C} \to \mathcal{D}$ assigns:

- To each object A in $\mathcal{C}$, an object $F(A)$ in $\mathcal{D}$
- To each morphism $f : A \to B$ in $\mathcal{C}$, a morphism $F(f) : F(A) \to F(B)$ in $\mathcal{D}$

Subject to two rules:

1. **Identity preservation:** $F(\mathrm{id}_A) = \mathrm{id}_{F(A)}$
2. **Composition preservation:** $F(g \circ f) = F(g) \circ F(f)$

Functors translate mathematical languages.
They map not just objects— but the arrows between them, while respecting structure.

Examples:
- The **forgetful functor** U : **Grp** $\to$ **Set** maps a group to its underlying set, and a group homomorphism to its underlying function.
- The **homology functor** in algebraic topology maps topological spaces to algebraic structures that capture holes and cycles.

Functors let us compare categories— by translating the behavior of one into another.
In category theory, functors are not just tools— they are first-class citizens.
If categories are worlds, functors are interdimensional bridges.

They carry not just forms, but meanings— preserving the logic of transformation across mathematical realms. A functor does not merely copy. It honors the rules of motion in a new domain.Like translating poetry without losing rhythm, a functor maps not only terms, but the grammar of relationships. **To define a functor is to witness one world through the structure of another.** This is the heart of abstraction: not to escape specifics, but to reveal the patterns that persist through change.
Where the arrow goes, and what it means to follow it— functors ensure the journey remains true.

20.3 Natural Transformations – Comparing Functors

Categories have objects and arrows. Functors map categories. But what maps functors?
Enter the **natural transformation**.

A **natural transformation** η between two functors $F, G : \mathcal{C} \to \mathcal{D}$ is a way of "smoothly" transitioning from F to G in a structure-preserving way.

Definition: A natural transformation $\eta : F \Rightarrow G$ assigns to each object A in $\mathcal{C}$ a morphism $\eta_A : F(A) \to G(A)$ in $\mathcal{D}$
Such that for every morphism $f : A \to B$ in $\mathcal{C}$, the following diagram commutes:

$$F(A) \xrightarrow{\eta_A} G(A)$$

$$\downarrow F(f) \qquad\qquad\qquad \downarrow G(f)$$

$$F(B) \xrightarrow{\eta_B} G(B)$$

This means:

$$\eta_B \circ F(f) = G(f) \circ \eta_A$$

Why it matters: - Natural transformations preserve the structure of transformations - They let us compare functors systematically - They form the morphisms in the **category of functors Analogy:** If functors are translations between languages, natural transformations are rules that ensure the translated meanings stay aligned. They are like dictionaries that not only match words—but maintain grammar and nuance. **Naturality** is the core idea: the transformation must respect the entire structure of morphisms. **It must be natural— not arbitrary, not forced, but flowing with the underlying logic of the category.** This is why category theory feels alive— it does not just name structures, it traces how those structures breathe through change.**A natural transformation is not just a bridge— it is a choreography of consistency.**It ensures that every path, every morphism, every transformation of objects is mirrored in meaning.When the diagram commutes, the universe aligns. Think of it like two alternate versions of the same story,translated not just word for word, but context for context,intention for intention— where every sentence in one version has a faithful echo in the other, through a lens of structured harmony. **Naturality is not about sameness. It is about coherence.**Not rigid, but rigorous.Not fixed, but faithful. Natural transformations are the grammar by which functors themselves converse. In the grand abstraction of mathematics, they whisper the deeper truth: *Not only can systems map to systems— but mappings themselves can be mapped.* This is the recursion of thought made formal. This is structure reflecting structure, not by force, but by nature.

20.4 Universal Properties – Defining by Uniqueness

In most branches of mathematics, we define an object by what it's made of.
In category theory, we define an object by what it does. More precisely, by how every other object relates to it.

universal property characterizes an object not by internal structure, but by a pattern of relationships that is unique up to isomorphism.

Example: The Product of Objects
In the category **Set**, the product of sets A and B is the set $A \times B$ with projections $\pi_1 : A \times B \to A$ and $\pi_2 : A \times B \to B$.
But we define it categorically by its universal property: For any object X with morphisms $f : X \to A$ and $g : X \to B$, there exists a unique morphism $\langle f, g \rangle : X \to A \times B$ such that:

$$\pi_1 \circ \langle f, g \rangle = f \pi_2 \circ \langle f, g \rangle = g$$

This uniqueness defines the product.

Other examples of universal properties:
- Terminal and initial objects - Pullbacks and pushouts - Limits and colimits - Free objects in algebra
- Adjunctions

Why it matters: - Universal properties abstract away construction - They focus on essence, not implementation - They let you recognize the same structure across different contexts **To define by univer-**

sal property is to define by role— not by recipe. In this light, category theory becomes the language of structural essence. But what is essence, if not the role something plays in the orchestra of all other things? A universal property is not an ingredient list. It is a spotlight— revealing what only that object can do in relation to the whole.It does not ask: "What is this made of?" It asks: "What unique position does this occupy within the web of morphisms?" **To define by universality is to let the world shape the thing— not the thing define the world.** In this way, category theory strips away decoration. It cares not for atoms or edges, but for the pattern that repeats, the form that returns regardless of substance.**When a structure plays the same role everywhere, it becomes canonical.**Not by decree. But by inevitability. This is the quiet power of universal properties: They render identity unnecessary. **To be defined by what only you can do, not what you are made of— that is the mathematics of meaning.** Each limit, each colimit, each initial and terminal object is a mirror in which every category sees itself more clearly. And that is what makes this abstraction not cold, but luminous.It does not dissolve difference. It

distills essence.**What remains, after all forms change, is what could never be replaced.**

20.5 Diagrams and Commutativity – Reasoning with Arrows

In traditional mathematics, we reason with symbols and equations.
In category theory, we reason with diagrams.

A **commutative diagram** is a picture where every path with the same start and end represents the same morphism.
It is not just an illustration— it is the proof itself.

Example: Commutative triangle

$$A[rd,"f"][rr,"h"]CB[ru,"g"]$$

This diagram commutes if:

$$h = g \circ f$$

That is, going from A to C directly via h is the same as going through B via f and g.

Why diagrams matter:
- They make reasoning visual and structural - They emphasize relationships over computation - They clarify universal properties and constructions

Large diagrams can express:
- Products, coproducts, pullbacks - Commuting squares, cubes, cones, and more - Entire theorems encoded in shapes **Proofs become paths. Concepts become**

flows. When a diagram commutes, you know the structure holds— without ever needing to expand equations. **Category theory turns reasoning into geometry.** And in that geometry, truth is not imposed— it is

followed.To follow a diagram is not to force an answer— but to trace what must be, given what already is. Each arrow is a commitment. Each node, a state of being. Each path, a journey through form. When two paths to the same place align, you have not proven a result— you have witnessed a necessity. **Commutativity is not agreement. It is inevitability.**This is the language of structure without calculation— a grammar of flow without coordinates. In the world of diagrams, you do not manipulate symbols. You walk along truths already paved by logic. Every triangle, every square, every cube— is a meditation on consistency. They do not simply display mathematics. They embody it.**Where paths converge, meaning becomes unambiguous.** And this is the deep grace of diagrammatic reasoning: You do not need to trust the equation— you see the structure.In the cathedral of category theory, diagrams are stained glass— they filter complexity into patterns your mind can walk through.And when they commute, you know that even among abstraction, there is a place where all ways lead home.

20.6 Category Theory as a Unifying Language

Most branches of mathematics have their own dialect. Set theory speaks in membership. Algebra in operations. Topology in neighborhoods. Logic in deduction. Computation in execution.
But category theory translates them all.

What does category theory unify?

- **Algebra:** Groups, rings, and fields become objects; homomorphisms become arrows.
- **Topology:** Spaces become objects; continuous maps are morphisms.
- **Logic:** Propositions become objects; proofs become morphisms.
- **Computation:** Types are objects; programs are morphisms.
- **Physics:** State spaces and transformations fit into categorical form.

Functors let us translate between categories.
Natural transformations let us compare those translations. **Universal properties** define by role, not substance. **Commutative diagrams** become the grammar of truth.

Category theory does not replace mathematics. It reveals its underlying architecture.

It shows that different fields are not separate towers— but views from the same structure. Mathematics becomes a network of transformations, rather than a pile of facts. And the study of mathematics itself becomes mathematical.**Category theory is not just a subject. It is a perspective.A mirror that mathematics holds to itself— and finally understands its form.** It is the mathematics of mathematics— a vantage point above the forest, where every branch reveals its place in the canopy of structure. Category theory does not ask: "What is this made of?" It asks: "How does this behave in relation to everything else?" **It shifts the question from substance to interaction, from parts to patterns.** Where others seek axioms, category theory seeks coherence. Where others prove existence, it reveals necessity. It does not erase detail— it abstracts meaning. It lifts mathematics from the particular to the inevitable. Like a river tracing the shape of the land, category theory reveals what cannot be otherwise.Each object exists not in isolation, but through its arrows—its connections, its maps, its relationships with all else. And so the shape of mathematics is not the sum of its parts, but the diagram that emerges when every function finds its domain.**It is the architecture of reason— not built from blocks, but from bridges.**To see through the lens of category is to see the soul of structure. And once you've seen it, you can never again believe that mathematics is a list of truths. You will know it is a single, unfolding relation.

Chapter 21

Infinity – Numbers Without Bound

Infinity is not a destination—
but a direction without end.
It is the breath between numbers,
the space that cannot close.
You do not arrive at it.
You unfold into it.

21.1 Potential and Actual Infinity

Infinity is not a number. It is an idea.
And like all deep ideas in mathematics, it wears many forms.

The first is **potential infinity** — a process that never ends.
Counting forward: $1, 2, 3, 4, \ldots$ You never reach the end, but you never stop.
This is the infinity of limits, of sequences, of "and so on."

The second is **actual infinity** — a completed totality.
The set of all natural numbers $\mathbb{N}$ is treated as a whole.
It has no last element. Yet it exists as a single object in set theory.

The distinction is ancient: Aristotle permitted potential infinity, but denied actual infinity.
Modern mathematics embraces both.
In calculus, we summon limits that never reach. In set theory, we define sets with infinitely many elements.

Infinity is not chaos. It is a rigorously defined frontier.
A landscape we can map, compare, and even measure.
To understand the infinite is to learn how structure persists when bounds disappear.

21.2 Cardinality – Comparing Sizes of the Infinite

Not all infinities are equal.
That is the astonishing discovery Georg Cantor gave the world.

The size of a set is called its **cardinality**.
For finite sets, cardinality is simple counting: the number of elements.
But what about infinite sets?

The set of natural numbers $\mathbb{N} = \{1, 2, 3, \dots\}$ is infinite.
So is the set of even numbers. So is the set of integers $\mathbb{Z}$.
Strangely, all three have the same cardinality.
Why?
Because they can be **put in one-to-one correspondence**.

Example:
Match each natural number to an even number:

$$1 \to 2, \quad 2 \to 4, \quad 3 \to 6, \quad \dots$$

Despite being "smaller," the evens are just as numerous as the naturals.

This shared size is called **countable infinity**, denoted $\aleph_0$ (aleph-null).

But some infinities are larger.

The set of real numbers $\mathbb{R}$ cannot be matched with $\mathbb{N}$.

Cantor's diagonal argument proves that any attempt to list all real numbers must leave some out.

Conclusion: The real numbers are **uncountably infinite**.

Their cardinality is strictly greater than $\aleph_0$.

This introduces an entire hierarchy of infinite sizes—some unimaginably vast.

Infinity is not a ceiling. It is a ladder.

21.3 Ordinal Numbers – The Order of the Infinite

Cardinality asks: *How many?* Ordinality asks: *In what order?*

An **ordinal number** is a way of arranging elements— even infinitely many— in a sequence.

For finite sets, ordinals and cardinals agree: 3 elements mean third position.
But in the infinite, order reveals a deeper layer.

The First Infinite Ordinal: ω (omega) is the order type of the natural numbers.
It represents: $0, 1, 2, 3, \ldots$ — an endless sequence with no last element.

Now add one more element after all of them:

$$0, 1, 2, 3, \ldots, \omega$$

This new order type is called $\omega + 1$.
Then:

$$\omega + 2,\ \omega + 3,\ \ldots,\ \omega \cdot 2,\ \omega^2,\ \omega^\omega,\ \ldots$$

Ordinal numbers stack up like exponents— infinitely tall structures with intricate branching.

Why it matters:

Ordinal numbers classify the structure of well-ordered sets. They model processes that extend beyond finite steps. They give form to transfinite induction and recursion.

Where cardinality counts, ordinality climbs.

Infinity is not just size. It is shape. It is sequence. It is process. not merely that which continues, but that which arrives.

Ordinality captures not just how much, but how far you've come.

It distinguishes between the never-ending staircase and the step you're standing on.

It sees difference between being forever in motion and having reached a new kind of beginning.

Each ordinal marks a stage not in length, but in unfolding.

Where cardinals weigh the crowd, ordinals line them up.

They teach us that infinity is not a blur— it is a choreography.

A dance of positions in the absence of end.

A story told not in mass, but in moment.

21.4 The Continuum Hypothesis – Gaps in the Infinite

Cantor showed us: there are more real numbers than natural numbers.
So the natural question followed:
Is there a size of infinity between $\aleph_0$ and the cardinality of the real numbers?
This is the **Continuum Hypothesis (CH).**
It asks:

$$\text{Is } \mathfrak{c} = \aleph_1?$$

Where $\mathfrak{c}$ is the cardinality of the continuum (the real numbers), and $\aleph_1$ is the next cardinal after $\aleph_0$.

In simple terms: Are the real numbers the smallest uncountable infinity?

The answer?
We cannot know.
In 1940, Kurt Gödel showed that CH cannot be disproven using the standard axioms of set theory (ZFC).
In 1963, Paul Cohen showed that CH cannot be proven either.

It is undecidable within ZFC.

This was a shock to mathematics: A question so central—so precise— has no answer unless we change the rules.

The infinite contains gaps of unknowability.
And we stand before them not with answers, but with awe.

Mathematics does not just prove. It learns where proof ends. The Continuum Hypothesis did not collapse under contradiction. It dissolved into ambiguity. It did not break the system. It revealed its boundary.

Mathematics met its horizon— not in chaos, but in silence.

To accept CH is a choice. To reject it is also a choice. Neither breaks the system. Both reshape its landscape.

This is not like deciding the value of $2 + 2$. This is like choosing a lens— and watching how the universe looks through it.

Set theory, once thought a bedrock, turns out to be terrain. Its axioms are not commands, but coordinates.

What does it mean to live in a world where not every question has an answer?

In physics, we expect uncertainty. In biology, we expect variation. But mathematics— was supposed to be certain.
And yet here, even in its purest form, we find freedom.

Different set theories spawn different worlds.

In one, $\mathfrak{c} = \aleph_1$. In another, $\mathfrak{c} = \aleph_{42}$.

These are not errors. They are different universes, each built on a consistent logic, each coherent within its own rules.

And so the continuum becomes not a bridge— but a question. Not a path— but a possibility.

The infinite is not one thing. It is many. Layered. Stacked. Unreachable. And perhaps unresolvable.

The Continuum Hypothesis invites us to ask: What happens when even mathematics must choose a belief? And it answers: Even in logic— freedom appears.

21.5 Infinity in Calculus, Physics, and the Real World

Infinity is not confined to theory.
It appears wherever reality resists finality.

In **calculus**, infinity defines the very core of motion and change.
Limits express what happens as a value approaches—but never reaches—a point. Derivatives and integrals rely on this idea, compressing infinite steps into finite meaning.

$$\lim_{x \to 0} \frac{\sin x}{x} = 1$$

An expression about the infinitely small— that becomes a tool of immense precision.

In **physics**, infinity is both a friend and a threat.
Black holes bend spacetime toward infinite curvature.
Quantum field theories wrestle with infinite energy densities— often requiring *renormalization* to subtract away the divergences.
The universe may be finite or infinite in extent— we do not yet know.

Even in computation, infinity appears as recursion, loops, or algorithmic undecidability.
Turing proved that some problems require infinite checking to solve— or can never be solved at all.

And in philosophy:
Infinity confronts us with questions about time, existence, the beginning of the cosmos, and the possibility of eternity.
It humbles the ego. It elevates the mind.
Infinity is not a number. It is a lens.
And through it, we glimpse the limit of what can ever be counted, proven, or known.

Yet for all its mystery, infinity is not abstract.
It is in the curve of a raindrop, modeled by
infinite tangents. It is in the light from distant stars, crossing infinite space to reach our eyes. It is in every digital signal, sampled from continuous waves that never truly end.

Mathematically, infinity marks where sequences point, not where they land. In real life, it marks what outpaces grasp, but not what escapes influence.

We cannot touch it— but we can measure its pull.

In Zeno's paradoxes, infinity tries to block motion. In calculus, we break that impasse with convergence. The sum of infinite steps becomes a finite distance. The impossible becomes traversable.

In modern cosmology, infinity haunts the start and the end. The singularity at the beginning. The potential heat death at the end. Two infinities—one dense, one empty. Both defying our current equations.

And in the fabric of our theories— from Hilbert's hotel to Gödel's incompleteness— infinity whispers: Not all things will fit. Not all truths can be captured. Not all structures can be completed.

Infinity is not the end of the story. It is the realization that the story has no end.

And yet— even without end, we keep going. Defining. Proving. Discovering.
As if the edge itself was not a wall, but a mirror, showing us who we are by how far we're willing to look.

Chapter 22

Chaos and Fractals – Order in Disorder

The edge of chaos is not the end of structure,
but its beginning—
where pattern hides inside unpredictability,
and the smallest nudge echoes forever.

22.1 Sensitivity to Initial Conditions – The Butterfly Effect

In classical mechanics, knowing the present meant predicting the future.
But chaos changed that.

Edward Lorenz, working on weather simulations in the 1960s, discovered that rounding a number from 0.506127 to 0.506 produced entirely different forecasts.
The system diverged rapidly, even though the starting conditions were nearly identical.

This gave rise to the idea of **sensitive dependence on initial conditions**.
Often called the **butterfly effect**— where the flap of a butterfly's wings in Brazil might set off a tornado in Texas.

In chaotic systems:

- Tiny changes grow exponentially
- Prediction becomes impossible in the long term
- Determinism does not guarantee predictability

Mathematically, this is often visualized using phase space diagrams, where nearby trajectories diverge wildly. **Chaos is not randomness.** It is deterministic unpredictability. The rules are exact. But the results are unstable. **Chaos reminds us: Even perfect**

knowledge can yield surprise. That surprise is not noise— it is

structure hidden too deep to grasp at once.What seems wild may still follow laws, but laws that amplify difference instead of erasing it.In chaotic systems, error doesn't stay small. It grows like a crack in glass, reaching across the whole.This changes what it means to understand. You don't just ask: "What happens?" You ask: "How fast do outcomes split?"Lyapunov exponents measure this divergence— the rate at which nearby paths fly apart.In weather, in ecosystems, in the beating heart and swinging pendulum— chaos lives.Not because there are no rules, but because the rules lead to rupture.And yet: there is a pattern in the storm. Fractals form. Attractors emerge.Chaos is not the end of knowledge. It is its turning point.It is the place where knowing must become humble, and precision becomes prophecy's limit. But chaos doesn't destroy understanding—

it demands a different kind. Not one of fixed prediction, but one of pattern, range, and expectation. In place of certainty, we track envelopes. We map strange attractors— not destinations, but gravitational forms toward which the system leans. We learn to ask not "Where will it go?" but "What kind of place will it orbit?" In chaotic systems, control shifts from exactness to containment. From precision to probabilistic shape.From path to landscape.This is why long-term weather prediction fails,but climate modeling endures. Not because we know tomorrow, but because we've mapped the range of
tomorrows.And in that range, order returns. Embedded in every fractal, in every feedback loop, is a law dressed in turbulence. Even when we cannot track the butterfly's flight,we can trace the space it flutters in. So chaos, like infinity, does not undo reason— it refines it.It says: You may not hold the future in your hand. But you can sketch the wind that shapes it.

22.2 Strange Attractors and Geometric Complexity

In most dynamical systems, repeated motion settles into something simple.
A pendulum slows into rest. A planet loops in an ellipse.
A heartbeat repeats in rhythm.
But in chaotic systems, the long-term behavior is complex— yet not random.
It tends toward an invisible shape called a **strange attractor**.

What is an attractor?
It's a set of states that the system approaches over time, regardless of starting point.
For simple systems:

- A point attractor (a stable equilibrium)
- A limit cycle (repeating orbit)

In chaotic systems, we find: **strange attractors**—fractal, non-repeating, but still bounded.

The most famous: **The Lorenz attractor**.
Its butterfly-wing shape maps the air currents of a weather model— never intersecting itself, never looping back— yet always staying within its wings.

Key features:

- Sensitive dependence
- Non-periodic but deterministic
- Fractal structure at every scale

Strange attractors prove that chaos contains order— but it is an order too complex to predict.

Geometry becomes the language of unpredictability. Patterns form not from rules alone, but from how rules unfold over time.

22.3 Fractals – Infinite Detail from Simple Rules

A coastline, a fern, a mountain ridge— each looks rough and irregular.
But zoom in, and the roughness repeats.
Smaller scales echo the whole.
This is the hallmark of a fractal.

A **fractal** is a shape that displays **self-similarity** across different scales.
Some are exact and mathematical, others are statistical and natural.

The Mandelbrot set, defined by the deceptively simple equation:

$$z_{n+1} = z_n^2 + c$$

produces one of the most complex, infinitely detailed structures ever visualized.
Zoom in, and more spirals, bulbs, and lightning bolts appear— each echoing the entire form.

Other famous fractals:

- **Koch snowflake** – Perimeter becomes infinite
- **Sierpiński triangle** – Triangles within triangles forever
- **Cantor set** – Infinitely many gaps in a finite line

Fractals appear in:

- Fluid dynamics
- Tree branching
- Market fluctuations
- Neuron growth
- Internet traffic

They show that complexity need not come from complexity. A few rules, iterated over time, can generate infinite structure.

Fractals reveal that nature is recursive, and scale is an illusion.

22.4 Chaos in Nature – Weather, Biology, and the Brain

Chaos is not rare. It is everywhere.
In the real world, perfect initial conditions never exist.
Noise is inevitable. The butterfly flaps every second.

Weather is perhaps the clearest example. Lorenz's discovery came from trying to model it.
Storms do not repeat. Clouds swirl without symmetry.
Yet they form and move within stable bands.
Weather is deterministic— but chaotic.

Biology also reveals chaotic dynamics:

- **Heartbeat rhythms** can become erratic yet structured— healthy variability lives near the edge of chaos.
- **Population models** like the logistic map show sudden jumps between stability and unpredictability.
- **Neural networks** in the brain oscillate and spike with fractal and chaotic patterns.

The brain, too, dances with chaos.
Consciousness emerges not from rigid order, but from dynamic systems— balancing between regularity and randomness.

Why does chaos appear in nature?
Because natural systems are:

- Highly sensitive to change
- Nonlinear in interaction
- Feedback-driven

Chaos is not failure. It is flexibility. It allows adaptation. It ensures diversity of outcome.
Life does not survive because it is predictable. It survives because it is chaotic— but contained.
Chaos is not the absence of pattern. It is pattern with a wild edge.

22.5 The Edge of Predictability – Chaos as a Creative Force

In the world of order, everything is known. In the world of randomness, nothing is useful. Chaos lives in between. This edge—between order and disorder— is where life happens, innovation occurs, and systems adapt.
Complex systems thrive here.

In physics: Turbulence in fluids exhibits chaotic flow—but this instability mixes energy, transports matter, and drives weather systems.

In biology: Gene networks do not follow strict logic gates. They pulse, oscillate, and spike— producing flexibility in development and repair.**In computation:** Artificial neural networks benefit from nonlinear, dynamic activation patterns. Too much order: overfitting. Too much randomness: no learning. Balance creates generalization.**In economics and culture:** Markets, ideas, and societies evolve at the edge of chaos— where novelty is neither suppressed nor uncontrolled.**Chaos is not what breaks the system.It is what makes the system breathe.** To control chaos is to kill it. To harness it is to understand its rhythms, its attractors, its recursive depths. **What seems wild may be woven with wisdom.** What seems unpredictable may be the only source of creativity.

Chapter 23

Mathematical Physics – The Code Beneath Reality

The universe is not written in words—
but in symbols, symmetries, and equations.
Its laws are not poetry, but precision.
To understand nature
is to uncover the syntax of spacetime.

23.1 Mathematics as the Language of Nature

Galileo wrote: "The book of nature is written in the language of mathematics."
Centuries later, we find it not metaphorical— but literal.

The laws of physics are mathematical structures. They are not just described by math. They are math.

- Newton's laws are equations of motion.
- Maxwell's equations govern electromagnetism.
- Schrödinger's equation drives quantum behavior.
- Einstein's field equations curve spacetime itself.

Each one reveals a layer of reality— and each one was discovered not by accident, but by following mathematical elegance. Why should this be?Why should something as abstract as a differential equation describe the movement of planets, the collapse of stars, the tick of a clock, or the orbit of electrons?Because mathematics, at its core, is the study of consistency. And reality, at its root, must be consistent to endure.**Physics is the consistency of the world.** Mathematics is the consistency of logic. Their union is not coincidence. It is necessity.**The universe is not just intelligible through mathematics. It may be mathematical in its very being.**

23.2 The Structure of Space and Time

We once believed that space was empty— a void to move through. And time was a constant drum— a rhythm outside the world. But mathematics told another story. In 1905, Einstein's theory of **special relativity** rewrote time.

$$E = mc^2$$

It unified mass and energy— and showed that time moves differently depending on how fast you move.
There is no universal now. Each observer has their own clock.

In 1915, his theory of **general relativity** bent space itself.

$$G_{\mu\nu} + \Lambda g_{\mu\nu} = \frac{8\pi G}{c^4} T_{\mu\nu}$$

This equation describes how matter and energy tell spacetime how to curve— and how curved spacetime tells matter how to move.The geometry of the universe is dynamic. Mass creates dents. Light bends. Orbits curve.Gravity is not a force. It is geometry.**Space and time are not containers.** They are participants— woven from the same mathematical fabric as the stars and particles they hold. **When math speaks of curvature, the universe listens.** And when spacetime ripples— as in gravitational waves— we hear math echo in the void.

23.3 Quantum Mathematics – Probability at the Core

In the quantum world, the equations remain precise— but their answers do not.
Particles don't follow simple paths. They interfere, superpose, tunnel. They blur.

Schrödinger's equation, the foundation of quantum mechanics, is:

$$i\hbar\frac{\partial}{\partial t}\Psi(x,t) = \hat{H}\Psi(x,t)$$

It governs the evolution of the **wave function** Ψ, which encodes all possible outcomes— but not which one will happen.

Measurement does not reveal what is. It selects from what could be.
Probability becomes fundamental. Not from ignorance, but from the structure of the world.

Feynman's path integrals suggest that particles explore all possible paths— and we observe the result of all paths interfering.

$$\text{Probability} \propto \left|\sum e^{iS/\hbar}\right|^2$$

Quantum mathematics teaches us:

- The world is not deterministic at its core.
- Uncertainty is not error—it is nature.
- Interference patterns are not anomalies—they are structure.

The more precisely you know a particle's position, the less precisely you can know its momentum.
This is not a flaw in instruments. It is a law of reality:

$$\Delta x \cdot \Delta p \geq \frac{\hbar}{2}$$

Mathematics reveals that beneath certainty lies structured uncertainty.
It is not chaos. It is quantum order— written in amplitude, phase, and complex number.

This is not a realm of what **is**, but of what **might be**—
a spectrum of weighted possibilities, each guided by
mathematical law, but realized only upon observation.
The equations do not dictate outcomes. They outline a canvas of interference— a probabilistic choreography, where all paths whisper their influence, and only one is made real through interaction.

Wave functions do not collapse because nature is indecisive— they collapse because reality is a selection from structure, a commitment born from entanglement, not authorship.

The mathematics of quantum theory does not describe a world beneath our senses. It **builds** that world— in tensors, Hilbert spaces, and eigenvalues— a realm where cause dissolves into configuration, and the future arises as a distribution waiting to be resolved.

What we call randomness is not the failure of reason, but the deeper success of math: the realization that form can generate unpredictability without ever becoming arbitrary.

Quantum mathematics is not mysticism. It is disciplined ambiguity. A logic not of yes or no, but of complex amplitudes, waiting for a question to define the answer.

23.4 Symmetry and Conservation – Noether's Theorem

In 1915, Emmy Noether proved a truth that reshaped the foundation of physics:

Every symmetry of a physical system corresponds to a conservation law.

This is **Noether's Theorem**— the hidden code beneath the laws we thought we understood.

Time symmetry (the idea that laws do not change over time) $\Rightarrow$ **Conservation of energy**

Spatial symmetry (the idea that laws are the same in every location) $\Rightarrow$ **Conservation of momentum**

Rotational symmetry $\Rightarrow$ **Conservation of angular momentum**

What does this mean?

That behind every quantity we thought was preserved— energy, momentum, charge— there is a deeper cause: a pattern that the universe respects.

Symmetry is not just aesthetic. It is generative.

The Standard Model of particle physics is built from gauge symmetries— abstract invariances that dictate how particles interact.

In mathematics: Symmetry becomes group theory, Lie algebras, and continuous transformations.

In physics: Symmetry becomes conservation, fields, and forces.

In reality: Symmetry becomes structure.

Noether showed that physics does not just describe what happens— it describes what must remain unchanged. **Symmetry is the fingerprint of necessity.** It is not merely what we see repeated— but what cannot be broken without breaking the universe itself. Noether revealed that invariance is law, not by declaration, but by consequence. That which stays the same gives rise to that which is conserved. This changes how we understand reality. We no longer ask only what forces act, but what symmetries are preserved. Why is energy conserved? Because time does not play favorites. Why is momentum conserved? Because space does not pick sides. From spin to charge, from quantum numbers to cosmic balance, every fundamental quantity is tied to a rule the universe will not break. And when symmetry is broken— mass emerges, particles differentiate, phase transitions carve the past from the possible.What we call change is often just a reorganization of what cannot change.In this light, physics is not about predicting motion. It is about discovering invariance— and seeing motion as the shadow cast by symmetry.

23.5 The Unreasonable Effectiveness of Mathematics

Why does the universe speak mathematics so fluently? It's not just that math describes nature. It often predicts nature— before observation ever confirms it.

Maxwell's equations predicted radio waves before they were detected.
Dirac's equation predicted antimatter before it was observed.
General relativity predicted black holes, gravitational lensing, and the expansion of the universe before we had the instruments to see them.

The physicist Eugene Wigner called this "**the unreasonable effectiveness of mathematics**."
Why should symbols invented in human minds align so perfectly with quarks and quasars?

Some believe mathematics is discovered— that it exists independently, and we simply uncover it.
Others believe it is invented— a tool that just happens to be useful.
But both views face a puzzle: usefulness beyond intention.
We do not set out to describe nature when we study pure group theory— yet it becomes the heart of particle physics.
We do not look for planets when working with differential geometry— yet it becomes spacetime.

Why does mathematics describe the world so well? Perhaps because the world *is* mathematics— a structure whose consistency can only be mirrored by logical form. **Or perhaps because our minds, shaped by that world, cannot help but speak its language.**

Either way— the mystery remains a portal, not a problem. **The universe does not resemble mathematics. It enacts it.**Not as a performer follows a script— but as structure sustains motion. As orbit follows curvature. As probability flows through amplitude. As light bends not from choice, but from the metric that frames its path. To say the universe enacts mathematics is not metaphor. It is the only way action holds form. When spacetime curves, it does so not with preference, but in exact

accordance with tensors. When particles decay, they do not ask permission— they follow group representations. Mathematics is not a tool we use to describe the cosmos. It is the pattern the cosmos follows before we arrive to describe it.Even our minds— even our symbols— are drawn from this deeper grammar. We are not observers standing outside the system. We are instances of the same coherence we study.To understand the universe, we do not look outward. We look inward, at the form that gave rise to the very act of asking.

Chapter 24

The Mathematics of Intelligence — Learning Machines

To learn is not to mimic.
It is to compress experience into expectation.
Intelligence is prediction refined by structure—
not because the system knows,
but because the world does.

24.1 Probability as Prediction

Before machines could recognize faces, generate sentences, or paint portraits from text, they had to learn something far more fundamental:

How to guess.

Not wildly. Not foolishly. But in the precise way that reality demands—not from certainty, but from structure. In a world where the wind might blow or not, where a cat might cross the road or sit still, intelligence does not emerge by waiting for clarity. It emerges by acting under ambiguity. The mind—biological or mechanical—must decide before it knows. It must place a wager on what the world will do next.

This wager is **probability**.

To a machine, a probability is not a feeling of doubt. It is a number. A ratio of belief to evidence. A bridge between what was and what might be.

The foundation of this bridge is an equation written centuries ago but reborn in every neural net today:

$$P(A|B) = \frac{P(B|A) \cdot P(A)}{P(B)}$$

Bayes' Theorem. A formula not just for adjusting numbers—but for learning.

When new evidence arrives, the machine updates its expectations. It reweighs its belief in a hypothesis. Not all at once, but gradually. Just as a child learns not from one lesson, but from many encounters. Probability becomes the tuning fork of understanding. Every new sound sharpens its pitch.

Conditional probability becomes the voice of experience. What is the chance that it will rain, given that the sky is dark? What is the chance that a word is "flower," given that the previous word was "petal"? Machine intelligence does not merely store outcomes—it tracks dependencies. Every connection adds resolution. Every correlation adds nuance. Over time, the system doesn't just predict—it anticipates.
It expects.

This is the first great shift from early logic to modern learning: **from certainty to likelihood.** Classical logic demands true or false. But intelligence lives in gradients. In doubt. In context.
Prediction, in the mathematical mind, is not about declaring the future. It is about preparing for it.

Even before a machine sees a sentence, it is already narrowing the space of what might come next.
If it sees the phrase "The sun will...", it silently weighs: rise, shine, set, burn. Each with a probability shaped by patterns seen before. It does not choose. It scores. And from these scores, it samples a possibility that makes the whole feel consistent.
This process is not magic.
It is math.

The **AI of today is not certain.** It is expectant. It does not know what you will type next, but it holds a weighted lattice of guesses. Each informed by prior data. Each shaped by structure. Each whispering, "This might be next."And when it speaks, it speaks not with authority, but with confidence proportional to pattern. The confidence, too, is mathematical. This is where machines diverge from old tales of thinking statues or rigid automata. The intelligence we build today does not emerge from rules carved in stone. It emerges from statistical learning—from the repeated folding of the past into the fabric of the future. Where humans saw randomness, machines see **inference**.
Where we once feared unpredictability, they learn to ride it—not by taming chaos, but by modeling its rhythm.

Probability, then, is not the opposite of truth.It is the grammar of learning. A grammar that turns experience into expectation.A grammar that writes the future in pencil.And it begins with a guess.
Not just any guess—but one that improves.

24.2 Representing Data in High-Dimensional Spaces

Before a machine can understand, it must first represent. In the world of artificial intelligence, this representation is rarely symbolic. It is geometric. The meaning of a word, image, or idea becomes a point—a vector—in a high-dimensional space, where distance encodes relationship and direction captures transformation. When we train a machine to learn, we are not handing it answers — we are shaping its internal geometry. Learning is not storage. It is spatial transformation.

Language as Tokens

Words are not treated as wholes. They are broken into **tokens** — subword units mapped to vectors in high-dimensional space.
Each token is embedded as a vector $x_i \in \mathbb{R}^d$. These embeddings evolve through the layers of the model, shaped by attention.

$$x_0 = \text{Embedding(token)} \Rightarrow x_L = \text{Final vector after } L \text{ layers}$$

Each token enters as a positionless presence—just a vector in space.
But as it passes through each layer, it is no longer alone. It begins to attend.

Self-attention reshapes every token vector, not by what it is, but by what surrounds it.
Each token asks: *"Who else matters to me?"*

And receives weighted answers from all others.
A new vector is formed— not from the token's identity, but from its relationships.

This is the key: *Meaning arises not from a token alone, but from its context.*
Each layer refines the vector using:

$$x_i^{(\ell+1)} = \text{Attention}(x_i^{(\ell)}, \{x_j^{(\ell)}\}) + \text{FeedForward}(x_i^{(\ell)})$$

Each step pulls the token closer to what matters, pushes it away from what doesn't.

- The early layers capture syntax: What is the verb? What is the object?
- The middle layers shape structure: Who is doing what? To whom? With what?
- The final layers reveal intent: What is implied? What comes next?

By the final layer, each token has traversed a path through the geometry of language. Its vector contains not just its own embedding, but the echo of every other token it attended to. This is not just embedding. This is evolution. Not just location in space— but transformation through attention.

Word Embeddings: Language as Geometry

Early neural networks treated words as one-hot vectors: sparse arrays with a single 1 and many 0s. These representations were isolated—each word orthogonal to the others, with no notion of similarity.Modern approaches, such as Word2Vec, GloVe, and FastText, learn continuous representations:

$$\vec{w}_{\text{word}} \in \mathbb{R}^d$$

where each word is mapped to a d-dimensional vector in a shared semantic space.The training objective often follows the **distributional hypothesis**—words used in similar contexts have similar meanings. Word2Vec, for instance, optimizes the following objective using Skip-Gram with Negative Sampling (SGNS):

$$\max \sum_{(w,c)\in D} \log \sigma(\vec{w} \cdot \vec{c}) + \sum_{(w,c')\in D'} \log \sigma(-\vec{w} \cdot \vec{c}')$$

Where:

- D is the set of observed word-context pairs.
- D' is the set of randomly sampled (noise) pairs.
- $\sigma(x) = \frac{1}{1+e^{-x}}$ is the sigmoid function.

The result: words that appear in similar neighborhoods are embedded nearby in vector space.These vectors are called **embeddings**. *Similarity becomes distance. Meaning becomes geometry.*

Sentence Embeddings — Context as Position

Words don't mean the same thing in every sentence. So models also build embeddings for entire phrases, sentences, even documents.Using transformers, each word attends to others — adjusting its vector in context:

$$\vec{w}_i' = \text{Attention}(\vec{w}_i, \{\vec{w}_j\}_{j \neq i})$$

Meaning is no longer static. It is positionally shaped.

Semantic Algebra: The Logic of Analogy

Once embedded, vectors allow language to be manipulated algebraically. A famous example:

$$\vec{v}_{\text{king}} - \vec{v}_{\text{man}} + \vec{v}_{\text{woman}} \approx \vec{v}_{\text{queen}}$$

This isn't trickery—it's structure. The model captures the relational transformation "male to female" and applies it across different semantic bases.This is not coincidence. It's algebraic analogy.This is not poetry. It is geometry. The angles, distances, and directions encode relationships — learned from billions of examples, shaped by optimization.These operations reflect consistent directions in vector space, where entire analogical structures become linear trajectories.

Dimensionality: What the Axes Mean

Embeddings exist in high dimensions—typically $d = 100$ to 300—but these dimensions aren't human-labeled. They emerge through optimization.
Each axis may correspond to subtle features:

- Formality or slang
- Positive or negative sentiment
- Technical vs colloquial use

The Geometry of Thought

What emerges is not just storage, but insight.

- Similarity becomes cosine similarity:

$$\cos(\theta) = \frac{\vec{u} \cdot \vec{v}}{||\vec{u}|| \, ||\vec{v}||}$$

- Clusters reveal categories.
- Distances imply confusion or clarity.
- Direction encodes transformation.

This is no longer just computation.
It is cognition—
expressed as shape. In this space:

- Clusters represent topics.
- Directions represent relations.
- Distances reveal semantic drift.

Visualized in 2D, embeddings look like clouds — diffuse, layered, curved. But in their native hundreds of dimensions, they are precise. **Language becomes a terrain. The model walks it by computation.**

Data as Points in Space

Every input — a sentence, an image, a sound — can be represented as a point in a high-dimensional space.

- A grayscale image of 28×28 pixels is a point in $\mathbb{R}^{784}$.
- A sentence can be embedded as a vector in $\mathbb{R}^d$ via word or sentence embeddings.

In this abstract space, similarity is spatial. Two similar items are **close together**. Two dissimilar ones are **far apart**. This is the core idea of **vector representations**. This is not just compression. It is alignment. Every point carries structure. Every distance encodes meaning.A cat and a tiger are closer than a cat and a toaster. Not because the machine understands "catness"— but because their features activate similar dimensions. The space becomes a landscape of meaning. Axes emerge not by design, but by usage. One direction might correspond to sentiment. Another to gender. Another to tense, or topic, or tone. These are not labeled by hand. They are inferred automatically from co-occurrence in data. The model does not learn words. It learns how words relate. It does not store knowledge. It positions knowledge into form. **Examples:**

- A facial recognition system doesn't memorize faces. It builds a space where your face lands near others that resemble you.

- A music recommendation engine doesn't know genres. It knows that the rhythm in one track echoes the rhythm in another.

- In image generation, a prompt becomes a point—and the system walks through space until it reaches a neighborhood where that concept lives.

This is how machines "understand." Not through language, but through geometry. Not by knowing what things are, but by placing them in relation.
In high-dimensional space:

- Ideas are shapes.

- Concepts are clouds.

- Clusters are meanings.

And between them—interpolation becomes imagination. This is the algebra of association. The topology of thought. And every layer of the neural network reshapes the terrain, so that separation becomes pattern, and pattern becomes prediction.

Representation Learning

Neural networks do not just classify. They transform. Each layer reshapes the data — rotating, stretching, bending — so that what was complex becomes simple in the final layer.In other words: learning is the search for a transformation that **untangles** the data.

In raw space, a cat and a dog may overlap.
In learned space, they lie on opposite sides of a decision boundary.

Beyond Words: Embedding Everything

The same logic applies across domains:

- **Image embeddings:** Convolutional Neural Networks (CNNs) transform raw pixels into vectors in $\mathbb{R}^d$.
- **Sentence embeddings:** Transformers (e.g. BERT, GPT) embed entire phrases, paragraphs, or documents.
- **User embeddings:** In recommendation systems, a user's behavior becomes a trajectory in preference space.

In each case, embedding models learn a mapping:

$$f : \text{Input} \to \mathbb{R}^d$$

Where proximity in $\mathbb{R}^d$ reflects similarity in meaning, behavior, or usage.

The Curse and Blessing of Dimensionality

High dimensions are strange. In $\mathbb{R}^2$, intuition serves us. In $\mathbb{R}^{10,000}$, it breaks.

- Distances become less meaningful — most points are equidistant.
- Volume concentrates in edges and corners.
- Local minima become rare — most minima are saddle points or flat regions.

But these very quirks become strengths:

- Models can separate highly complex patterns.
- Interpolation becomes smoother.
- Representations can disentangle abstract structure.

In spaces with many dimensions:

- Volume increases exponentially — data becomes sparse.
- Distances lose meaning — all points are nearly equidistant.
- Visualization becomes impossible — intuition collapses.

AI models often operate in spaces with 512 or 1024 dimensions. Reducing this to 2D or 3D is not just for illustration — it's for insight.

The Geometry of Knowledge

Modern large language models (LLMs) operate over these embedded representations. They don't "know" meanings in the human sense. They inhabit a vector field sculpted by usage, where concepts ripple and flow.
Learning becomes the shaping of this field.
Prediction becomes the traversal of it.
Understanding becomes the mapping of trajectories through it.
This is the geometry of meaning — the quiet coordinates beneath intelligence.

To model the mind, we mapped the meaning. And in that map, machines began to think.

Meaning Without Definitions

No dictionary is needed. Just enough text.
As models see more examples, their embedding space reshapes. Words migrate. Ideas cluster. Analogies form.
The machine never defines. It aligns.
Embeddings are not about truth. They are about proximity.

Meaning is where you are in relation to everything else.

Dimensionality Reduction — Seeing the Shape of Thought

Modern AI operates in vast, invisible spaces.
Each input — an image, a sentence, a sound — is transformed into a point in a space with hundreds or thousands of dimensions. Each dimension may capture a nuance: color gradient, grammatical tense, or the presence of a curve in a digit.
But high-dimensional spaces are hard to understand. To interpret how the system thinks, we must reduce its thoughts to something we can see.
This is the purpose of **dimensionality reduction.**
Dimensionality reduction techniques like PCA (Principal Component Analysis) or t-SNE help visualize these embeddings:

$$\text{PCA: } \vec{y_i} = W^\top(\vec{x}_i - \vec{\mu})$$

Where:

- $\vec{x}_i$ is the original vector.
- $\vec{\mu}$ is the mean vector.
- W contains the top eigenvectors of the covariance matrix of the data.

This reveals that the geometry of meaning is not arbitrary—it is compressed structure, learned from data.
Principal Component Analysis (PCA) finds directions (components) that capture the most variance in the data.

Given a dataset X, PCA finds
an orthogonal basis $\{v_1, v_2, ..., v_k\}$ such that:

$$v_1 = \arg\max_{\|v\|=1} \mathrm{Var}(Xv)$$

The data is then projected:

$$X_{\text{reduced}} = X \cdot [v_1, v_2, ..., v_k]$$

- PCA preserves *linear* structure.
- It is fast and interpretable.
- But it cannot capture complex, curved manifolds.

t-Distributed Stochastic Neighbor Embedding (t-SNE) focuses on preserving local structure.
It builds two probability distributions:

- P_{ij} — Similarity in high-dimensional space.
- Q_{ij} — Similarity in low-dimensional space.

It minimizes the Kullback–Leibler divergence:

$$\mathrm{KL}(P||Q) = \sum_{i \neq j} P_{ij} \log \frac{P_{ij}}{Q_{ij}}$$

- Points that were neighbors stay neighbors.
- Great for revealing clusters and categories.
- But can distort global geometry.

UMAP builds on topological ideas.
It:

- Constructs a fuzzy topological structure in high dimensions.
- Optimizes a layout in low dimensions that preserves this structure.

UMAP better captures both local and global relationships. It is increasingly used to visualize neural networks and embeddings.
Dimensionality reduction doesn't just shrink data. It reveals structure.
In language models:

- Words become vectors in high-dimensional space.
- Similar words cluster — "king" and "queen" are close.

These embeddings are not programmed. They are learned — as the system trains.
Concepts become directions. Meaning becomes geometry.
We reduce dimensions not to lose complexity — but to see it.
We do not flatten thought. We translate it.
When we project the invisible into view, we glimpse the mind of the machine.
Not through its logic — but through its landscape.

Dimensionality reduction reveals the contours of cognition.

24.3 Neural Networks and Transformers

A neural network is not a brain. But it is inspired by one. It is a system of functions layered together, where each layer transforms the input into a slightly more useful form. What emerges is a hierarchy of abstractions — built not by intention, but by propagation.

At the heart of modern artificial intelligence is a system built from nothing more than weighted sums and nonlinear twists.

Neural networks are not brains. They are layered algebra machines — but ones that can learn to approximate anything.

The Structure of a Neural Network

A basic neural network consists of:

- **Input Layer:** Receives raw data (e.g., pixel values, numerical vectors).
- **Hidden Layers:** Intermediate layers where computation happens. These layers extract features from input data.
- **Output Layer:** Produces the final prediction (e.g., a classification label or numerical value).

Each layer performs a linear transformation followed by a non-linear function:

$$z^{(l)} = W^{(l)}x^{(l-1)} + b^{(l)}$$

$$a^{(l)} = f(z^{(l)})$$

Where:

- $W^{(l)}$ is the weight matrix of layer l,
- $b^{(l)}$ is the bias vector,
- f is the activation function (e.g., ReLU, sigmoid, tanh),
- $x^{(l-1)}$ is the input from the previous layer.

Each neuron performs:

$$z = w_1x_1 + w_2x_2 + \cdots + w_nx_n + b$$

$$a = \sigma(z)$$

Where:

- x_i are inputs
- w_i are weights
- b is bias
- σ is a nonlinear activation (like ReLU or sigmoid)

The output of one layer becomes the input to the next.

Activation Functions — Injecting Nonlinearity

Without non-linear activation
functions, neural networks would collapse into a single linear transformation — no matter how many layers are

stacked. Activation functions allow networks to model complex, non-linear
patterns.
Common choices include:

- **ReLU (Rectified Linear Unit):** $f(x) = \max(0, x)$
- **Sigmoid:** $f(x) = \frac{1}{1+e^{-x}}$
- **Tanh:** $f(x) = \tanh(x) = \frac{e^x - e^{-x}}{e^x + e^{-x}}$

Each function warps the signal, preserving some elements and muting others — shaping how information flows through the network.

Depth and Hierarchy — Learning in Levels

The power of deep learning comes from depth: multiple layers learning sequential transformations. Early layers capture low-level features (edges, colors). Later layers combine them into high-level abstractions (shapes, objects, meanings).
In image recognition:

- Layer 1: Edge detection
- Layer 2: Patterns and textures
- Layer 3: Object parts
- Layer 4+: Whole object recognition

In language models:

- Layer 1: Character or word embeddings

- Layer 2: Grammatical structure
- Layer 3+: Semantic relationships, logic, style

This layering is not programmed. It is discovered — an emergent shape carved by optimization.
More layers mean more representational power.

- Shallow networks can approximate basic functions.
- Deep networks can represent hierarchies — edges to shapes to concepts.
- Convolutional networks (CNNs) specialize in vision.
- Recurrent networks (RNNs) and Transformers in language.

Each layer refines meaning. Each transformation is a new coordinate system in the space of thought.

The Universal Approximation Theorem

Mathematically, neural networks are universal function approximators. Given enough neurons and the right weights, a network can approximate any continuous function on a compact domain:

$$\forall \epsilon > 0, \exists\, f_\theta \text{ such that } |f(x) - f_\theta(x)| < \epsilon$$

This does not guarantee efficiency — just possibility. The map exists, but the journey is hard.

Decision Boundaries

At the end of a neural network, a decision must be made. This is typically done using a **softmax** function for classification:

$$\sigma(z_i) = \frac{e^{z_i}}{\sum_{j=1}^{K} e^{z_j}}$$

This transforms the network's outputs into a probability distribution over K classes.

But underneath this final layer lies a geometric truth: the network has reshaped the space so that each class occupies a distinct region, and the boundaries between regions are **hyperplanes** in $\mathbb{R}^n$.

When Functions Begin to See

A trained neural network does not know. It maps. It learns associations, patterns, and transitions. But when those patterns hold across unseen data — it appears to understand.

What you're seeing is mathematics performing pattern recognition in high-dimensional space. Each neuron is not a thought — it is a signal transformer.

But the system, as a whole, begins to exhibit surprising behavior: abstraction, generalization, even creativity.

That is the mathematics of emergence — and we are only beginning to chart its laws.

Attention and Transformers — The Rise of Language Models

Not all inputs matter equally.
In a sentence, some words carry weight. Others point, redirect, or amplify meaning. Language is a web of dependencies — long-range, subtle, contextual.
Traditional models treated words like fixed neighbors. But intelligence in language requires something else: *Attention.*
Before transformers, language models read sequentially — word by word, in order. But understanding does not depend on order alone. It depends on relationships.
Transformers changed the field by introducing *attention* — a mechanism that lets models look at all parts of a sentence simultaneously, weighting what matters most.
Introduced in 2017 ("Attention Is All You Need"), the transformer architecture changed AI.
Its core innovation: **attention**.
Instead of reading data sequentially, it looks everywhere at once — calculating how each token relates to every other:

$$\text{Attention}(Q, K, V) = \text{softmax}\left(\frac{QK^\top}{\sqrt{d_k}}\right) V$$

Where:

- Q = Queries
- K = Keys
- V = Values

- d_k = Dimension of key vectors

Each word is transformed into three vectors:

- **Query (Q)** — What the word is asking for
- **Key (K)** — What the other words offer
- **Value (V)** — What information is carried

- QK^T: Measures similarity between query and key
- $\sqrt{d_k}$: Scales scores to avoid exploding values
- Softmax: Converts scores into probabilities
- V: Aggregates the information using attention weights

The result: each word's meaning becomes a function of its context — not just its position.
Each word becomes a mixture of others — shaped by relevance, not just sequence.
Attention allows the model to weigh relationships dynamically:

- "The cat sat on the **mat**" — attention links "sat" to "cat."
- Long-range dependencies become learnable.

Transformers power modern LLMs like GPT and BERT. They don't just process text. They *model dependencies*, *anticipate structure*, and *internalize syntax*.
Each block consists of:

- Multi-head self-attention

- Feed-forward network
- Layer normalization and residual connections

Instead of one attention computation, transformers use several in parallel:

$$\text{MultiHead}(Q, K, V) = \text{Concat}(\text{head}_1, \ldots, \text{head}_h)W^O$$

Each **head** learns a different way to focus:

- Syntax (e.g., subject-verb pairs)
- Semantics (e.g., who did what)
- Long-range references (e.g., pronouns)

This parallelism allows transformers to detect structure and nuance — without explicit rules.
Positional encodings are added to word embeddings to encode sequence order.

$$\text{Embedding} = \text{Token Vector} + \text{Positional Vector}$$

Since transformers process all words at once, they need a way to know order. They add a *positional encoding* to each word vector:

$$\text{PE}_{(pos,2i)} = \sin\left(\frac{pos}{10000^{2i/d_{model}}}\right) \quad \text{PE}_{(pos,2i+1)} = \cos\left(\frac{pos}{10000^{2i/d_{model}}}\right)$$

This encodes position as a wave — letting the model sense relative distances mathematically.
This allows transformers to process sequences in parallel while preserving order.

After attention, transformers apply a feedforward network to each word:

$$\text{FFN}(x) = \max(0, xW_1 + b_1)W_2 + b_2$$

This reshapes the information — adding depth and abstraction.

One transformer layer is powerful. Dozens, stacked, become profound.

Each layer builds on the last — refining attention, sharpening meaning, growing capacity. From literal words to abstract patterns, each level captures new structure.

The final layer produces a probability distribution over the next token:

$$P(\text{token}_{t+1}) = \text{softmax}(Wx_L + b)$$

Language models are not programmed with grammar. They are trained to predict:

$$P(w_t | w_1, w_2, \ldots, w_{t-1})$$

This is **next-token prediction** — the foundational task of GPT, BERT (masked token prediction), and other LLMs.

Over billions of examples, the model learns the patterns of language — not rules, but statistical structure.

It predicts what makes sense, not because it understands, but because it fits.

Large-scale transformers, like GPT, contain:

- Hundreds of layers

- Billions of parameters
- Trained on terabytes of text

They perform zero-shot, few-shot, and chain-of-thought reasoning not because they were taught *how*, but because patterns of human reasoning are embedded in the training data.

Transformers are not "smart." They are mathematical structures tuned to minimize loss over structured sequences.

They cannot think.

But they can propagate thought-like behavior.

And the math behind them is not magic. It is:

- Dot products
- Linear algebra
- Probability distributions
- Optimization landscapes

Mathematics does not generate meaning. It generates structure that can host it.

- Contextual meaning: The same word shifts depending on others.
- Long-range coherence: A sentence connects to previous ones.
- Emergent knowledge: Models discover syntax, analogy, and inference.

Transformers do not read. They align. Not with rules — but with relationships.
They don't see grammar. They see patterns. Not because they're taught — but because they optimize prediction.

Meaning emerges not from instruction, but from compression.

Transformers are not programs. They are patterns — encoded in weights, trained on text, shaped by entropy. They don't "understand." They approximate — in millions of dimensions.
And yet: from this blur, meaning arises.

From Layers to Language

A trained neural network is a geometry of meaning. Each layer is a transformation. Each weight is a memory. Each output is a projection from high-dimensional space to human-readable form. The network doesn't *understand.* But it maps structure to signal — thought to form.

Neural Algebra

Though inspired by the brain, these systems are purely mathematical. They are:

- Compositions of linear algebra and calculus.
- Guided by loss functions and shaped by optimization.
- Capable of abstracting — not because they feel, but because the math aligns.

A neural network is not alive. But it is structured to survive — in the environment of data.It is an equation that rewrites itself. And in that rewriting, it becomes a mirror — not of us, but of the patterns we leave behind.

24.4 Optimization and Learning

At the heart of machine learning lies a single, recurring question:
How do we improve?
Neural networks learn not by memorizing data, but by optimizing a function — a mathematical landscape of error — and slowly adjusting their internal parameters to move toward less error.
At the heart of learning lies a single goal: minimize the distance between what the model expects and what the world delivers. This is the domain of optimization — where algorithms refine themselves to reduce error.
Prediction is not enough. To improve, a system must search. Optimization is the process of finding the model parameters that minimize the loss — of navigating the landscape of error to reach a valley of usefulness.
A machine does not understand. It adjusts.
Each "thought" it has — each output it offers — is judged not for truth, but for error. And that error flows backward.

The Learning Loop

At the core of machine learning lies a loop:

1. Input is passed through the network: $x \to \hat{y}$
2. The output $\hat{y}$ is compared with the true label y
3. A loss is calculated: $\mathcal{L}(\hat{y}, y)$
4. Gradients are computed: how much each weight influenced the error
5. Each weight is adjusted: nudged to reduce future error

This loop repeats thousands — millions — of times. The model does not acquire "knowledge." It minimizes error.

Loss Functions — Measuring the Miss

To learn, a system must know how wrong it is.
A **loss function** quantifies the error between a model's prediction and the actual outcome. It is the feedback loop that transforms ignorance into structure.
Every neural network — every intelligent model — is a sculptor of approximations. But without a chisel, the form cannot emerge.
The loss function is that chisel:

- It penalizes mistakes.
- It provides a direction for correction.
- It defines what "good" means in a mathematical landscape.

Let θ represent the parameters of a model, x the input, and y the desired output. The model's prediction is $f_\theta(x)$.

A **loss function** $L(f_\theta(x), y)$ quantifies how far the prediction is from the truth.
If the model guesses 0.7 and the correct answer is 1, the loss might be:

$$\text{Loss} = (y - \hat{y})^2$$

Common loss functions:

- **Mean Squared Error (MSE):** Used in regression problems to penalize large deviations.

$$L = \frac{1}{n}\sum_{i=1}^{n}(y_i - \hat{y}_i)^2$$

- **Cross-Entropy Loss:** Used in classification. Measures the distance between two probability distributions.

$$L = -\sum_{i=1}^{n} y_i \log(\hat{y}_i)$$

- **Hinge Loss:** Common in support vector machines. Encourages correct classification with margin.

$$L = \sum_{i=1}^{n} \max(0, 1 - y_i \cdot \hat{y}_i)$$

Each loss function encodes an assumption about the problem. It defines what it means to be close — or catastrophically wrong.
Loss functions are not just numbers. They are landscapes.

- **Flat terrain** means many weight combinations give

similar outcomes.

- **Sharp valleys** mean the model is sensitive to small changes.
- **Local minima** can trap learning.

The shape of the loss surface determines how hard it is to learn.
Sometimes, the model fits the data too well — memorizing noise rather than generalizing structure.
Regularization terms are added to the loss function to discourage complexity:

- **L1 regularization:** Promotes sparsity (many weights = 0)
$$L' = L + \lambda \sum |w|$$
- **L2 regularization:** Penalizes large weights (smooths the model)
$$L' = L + \lambda \sum w^2$$

Here, λ is the regularization strength — a knob tuning humility into the system.
A model is only as good as the signal it is punished for missing.
Define the wrong problem — get the wrong intelligence.
The loss function, then, is not a mere tool. It is the ethical spine of the algorithm.
It teaches the machine what to value — and what to ignore.
What you penalize, you shape. What you reward, you amplify.

This is true not just in machine learning — but in every system that learns, adapts, or evolves.
Loss is not failure. It is the engine of correction.
These functions are gradients' guideposts. They shape the path through the landscape of weights and possibilities.

Backpropagation — The Chain Rule of Learning

Neural networks are layered — each layer transforms the input before passing it on.
To adjust the parameters of all layers, we apply the chain rule of calculus:

$$\frac{\partial L}{\partial \theta_i} = \frac{\partial L}{\partial a_n} \cdot \frac{\partial a_n}{\partial a_{n-1}} \cdot \dots \cdot \frac{\partial a_{i+1}}{\partial \theta_i}$$

This algorithm — called **backpropagation** — flows the gradient backward from output to input.
Every connection learns how it contributed to the error — and how to adjust.
Neural networks use **backpropagation** to spread the error signal backward through the layers. This process calculates how each parameter contributed to the final error and nudges it accordingly:

$$\delta^{(l)} = \left(W^{(l+1)}\right)^T \delta^{(l+1)} \circ f'\left(z^{(l)}\right)$$

Where:

- $\delta^{(l)}$ is the error at layer l,
- $W^{(l+1)}$ is the weight matrix of the next layer,

- $f'\left(z^{(l)}\right)$ is the derivative of the activation function.

Through this recursive chain rule, models learn which parts of themselves need correction.
Let a simple neural network layer be:

$$z = Wx + b, \quad a = \sigma(z)$$

To update the weights W, we must compute:

$$\frac{\partial \mathcal{L}}{\partial W} = \frac{\partial \mathcal{L}}{\partial a} \cdot \frac{\partial a}{\partial z} \cdot \frac{\partial z}{\partial W}$$

Each term is passed backward — from output to input — multiplying along the way. This is the gradient.
The algorithm proceeds layer by layer, adjusting each weight based on its contribution to the error.
The network learns by adjusting its weights to minimize a loss function L — a measure of error.
Weights are updated using **gradient descent**:

$$w \leftarrow w - \eta \frac{\partial L}{\partial w}$$

This is not memorization. It is *pattern adjustment* — tuning the surface of a multidimensional function.

Gradient Descent — Learning by Slopes

Learning in modern AI is not memorization. It is movement — a descent through a landscape shaped by error.
Gradient descent is the core algorithm behind how machines learn. It doesn't tell the model what to think. It shows the direction to move — to think a little less

wrongly.
To improve, we compute the gradient of the loss function:

$$\nabla_\theta L$$

This vector tells us the direction of steepest ascent in error. To minimize, we go the opposite way:

$$\theta \leftarrow \theta - \eta \nabla_\theta L$$

Where:

- θ are the parameters,
- η is the learning rate — how big a step we take,
- $\nabla_\theta L$ is the gradient of the loss function with respect to the parameters.

Each update moves the model toward less error.
This is called **gradient descent.**
Most modern AI systems use variants of **gradient descent** to optimize parameters. The idea is simple, yet powerful:

- Calculate the gradient (slope) of the loss with respect to each parameter.
- Move in the opposite direction of the gradient.
- Repeat until improvement stalls or a minimum is reached.

High error pulls the model uphill. Gradient descent rolls it down.

Mathematically, the gradient is the vector of partial derivatives:

$$\nabla L = \left[\frac{\partial L}{\partial w_1}, \frac{\partial L}{\partial w_2}, \ldots, \frac{\partial L}{\partial w_n}\right]$$

Each component tells how much a small change in a parameter increases or decreases the loss.
In practice, full gradients are expensive to compute. So we use **Stochastic Gradient Descent (SGD)**:

$$w \leftarrow w - \eta \cdot \nabla L_{\text{batch}}$$

Instead of using all data, the model samples mini-batches — small random subsets.
This introduces noise into the process — but also helps escape bad minima and speeds up convergence.
Variants:

- **Stochastic Gradient Descent (SGD):** Uses a single or small batch of data points per update, introducing noise but increasing efficiency.

- **Mini-Batch Gradient Descent:** A compromise between full-batch and stochastic, balancing speed and stability.

- **Momentum:** Adds velocity to updates to avoid getting stuck in local minima. Accelerates learning by smoothing the gradient using a running average.

$$v_{t+1} = \gamma v_t + \eta \cdot \nabla_\theta L(\theta_t)$$

$$\theta_{t+1} = \theta_t - v_{t+1}$$

- **Adam:** Combines momentum and adaptive learning rates to converge faster. Maintains estimates of first and second moments of the gradient:

$$m_t = \beta_1 m_{t-1} + (1 - \beta_1)\nabla L_t$$

$$v_t = \beta_2 v_{t-1} + (1 - \beta_2)(\nabla L_t)^2$$

$$\hat{m}_t = \frac{m_t}{1 - \beta_1^t}, \quad \hat{v}_t = \frac{v_t}{1 - \beta_2^t}$$

$$w_t = w_{t-1} - \eta \cdot \frac{\hat{m}_t}{\sqrt{\hat{v}_t} + \epsilon}$$

Adam adjusts each parameter's learning rate based on its history — helping models converge quickly and accurately.
Modern optimization methods — like Adam, RMSprop, and momentum-based descent —
adjust learning dynamically
to escape such traps.
Each optimizer defines a different philosophy of progress — how boldly or cautiously the model explores its space of solutions.

The Learning Landscape

Every model has a *loss landscape* — a mathematical surface shaped by its parameters. Learning is navigating this landscape toward a minimum. But this surface is rugged:

- **Local minima** may trap the model in suboptimal solutions.

- **Saddle points** are flat regions where gradients vanish, slowing progress.
- **Global minima** represent the best possible solution, often hard to find in high-dimensional space.

The loss function is not flat. It is a complex surface — with hills, valleys, ridges, and basins.

- **Local minima:** Small dips in error — not always optimal
- **Saddle points:** Flat but unstable — neither up nor down
- **Global minimum:** The best possible configuration

Deep networks have millions of parameters. The landscape is not two-dimensional — it's a manifold in hundreds of thousands of dimensions.
Gradient descent is like rolling a marble across this terrain — always downhill — hoping that the bottom you find is good enough.

- Flat regions: slow learning (small gradients)
- Sharp valleys: unstable updates

Optimization is rarely a smooth downhill ride.
The landscape of loss is:

- Full of **local minima** — valleys that are not the deepest.

- Interrupted by **saddle points** — flat, ambiguous regions that delay learning.
- Littered with **plateaus** — where gradients vanish and progress stalls.

Modern deep learning doesn't seek a global minimum — it seeks a *useful* one. A wide, stable valley that generalizes well.

Optimization transforms a model's structure. But the shape of the solution depends not just on data — but on the terrain created by the loss function and parameter space.

A model is not trained in isolation. It is shaped by:

- The curvature of the loss surface.
- The dimensionality of the parameter space.
- The constraints imposed by regularization.

The optimizer is not just a mechanic. It is an explorer — mapping the terrain of possible minds.

Modern architectures rely on:

- **Smooth activations** (like ReLU)
- **Normalized inputs**
- **Smart initializations**

These help navigate the landscape efficiently.

Beyond Gradients — Evolution, Swarms, and Sampling

Not all intelligence learns by slopes.
Some systems use alternative optimization strategies:

- **Genetic Algorithms:** Inspired by evolution — mutations, selection, recombination.
- **Particle Swarm Optimization:** Agents explore the space by sharing discoveries.
- **Markov Chain Monte Carlo (MCMC):** Uses probability to sample from complex distributions.

These approaches are slower — but powerful when gradients are unavailable, unreliable, or misleading.

No Teacher — Only Feedback

There is no ghost in the model. No overseer. No teacher explaining concepts.
There is only feedback.
The model outputs. The system judges. The error flows.
The model adapts.
This is not comprehension. This is adjustment.
The illusion of understanding comes from optimization under structure.

When Feedback Fails

Models do not learn truth. They learn to minimize a number.

If the loss function misaligns with what we value — fairness, reasoning, explanation — the system optimizes the wrong thing.

Feedback drives behavior. But feedback must be meaningful.

In deep learning, error is a compass. But a compass only helps if the map reflects the world.

A Learning Machine

Learning is not magic. It is calculus applied recursively. It is error, feedback, and structure aligning through thousands of iterations. What looks like intuition is math sculpting itself in silence — until a pattern begins to hold. Learning is not magic. It is iteration.
A neural network:

1. Makes a guess.
2. Measures the error.
3. Adjusts to reduce the mistake.
4. Repeats.

Over millions of updates, a pattern emerges — not because the model understands, but because it optimizes. **Intelligence, in machines, is structured trial and error.**
Mathematics does not feel — but it learns.
The model does not know truth. It minimizes error.

It cannot explain why a sentence works — but it can predict what comes next.
Each descent step changes billions of values. And yet, together, they form understanding.
Gradient descent is not thought. It is pressure. Not wisdom. But adaptation.
And from this — structure arises.

Learning is not magic. It is alignment with patterns.

Optimization as a Philosophy

The choice of optimizer is more than technical. It reveals a worldview:

- Should intelligence chase every slope? Or wander widely?
- Should it trust momentum? Or adapt on the fly?
- Should it seek perfection? Or settle into stability?

Optimization is how we encode bias, balance, and belief. It is the mathematics of improvement — and the silent sculptor of systems that learn.

24.5 Scaling Laws — The Mathematics of Bigger, Better, and Beyond

More data. More parameters. More compute.
These are not just brute-force increases — they follow a law. A law discovered not by decree, but by empirical clarity:
Performance improves predictably as models grow.

The Scaling Law Curve

Let L represent the loss (error) of a model, and let N, D, and C represent:

- N: number of parameters
- D: size of dataset
- C: compute used in training

Empirical studies show:

$$L(N) \propto N^{-\alpha}, \quad L(D) \propto D^{-\beta}, \quad L(C) \propto C^{-\gamma}$$

Where α, β, and γ are positive constants that vary by task and architecture — typically between 0.05 and 0.1. This means that as model size increases exponentially, the loss decreases along a smooth power-law curve.
Bigger really is better — up to a point.

The Regimes of Growth

Scaling obeys three intertwined laws:

1. **Model Scaling:** Larger neural networks can fit more complex functions.

2. **Data Scaling:** More training data prevents overfitting and reveals deeper patterns.

3. **Compute Scaling:** More compute enables longer, deeper, better training runs.

When all three increase together, performance improves predictably.
This relationship is not magic — it is the geometry of optimization in high-dimensional parameter spaces.

Diminishing Returns — But Not Randomly

Though each doubling in model size yields less absolute improvement, the gains are consistent.
Loss decreases log-linearly with compute and scale:

$$\log L \sim -a \log C + b$$

This predictability gives rise to an astonishing possibility: **Extrapolation.** We can forecast model performance years in advance — not through speculation, but through the continuation of a curve.

Chinchilla and the Data/Compute Trade-off

Early models like GPT-3 used more parameters than data.
Later models, like Chinchilla (DeepMind, 2022), optimized the ratio between model size and training tokens, showing that:
Smaller models trained on more data can outperform larger ones with less training.
This led to a new insight:

$$\text{Optimal compute budget} \quad \Rightarrow \quad N \propto D^{0.73}$$

Where N is the number of parameters and D is the size of the dataset.

This relationship means there is an *optimal tradeoff* between how big a model should be and how much data it should see.

- Too large a model on too little data? It memorizes noise.
- Too much data for a small model? It lacks the capacity to absorb it.

But tuned correctly, a model becomes efficient—both in training and inference.

Chinchilla showed:

- Training tokens matter more than we thought.

- Model size without sufficient data leads to underfitting.
- Efficient architectures require matched compute, not maximal compute.

This shifted how the AI community measures cost:

- From FLOPs-per-parameter → to FLOPs-per-token
- From model bloat → to training-efficiency curves

The new goal is not to be *larger*—but to be *right-sized.*

And in this insight, the field took a leap:
From maximalism to balance.
From size wars to science.

Scaling is not Guesswork

At the heart of scaling laws is the recognition that learning in artificial systems is not an art — it is a measurable curve.
Math does not just describe behavior. It predicts the future of learning itself.
We now know:

- How model performance grows with size
- How to balance model, data, and compute
- How to anticipate limits and breakthroughs

This transforms machine learning from intuition into engineering — and from mystery into mathematics.

The frontier is now governed by curves. The frontier is now governed by curves. Not lines of code— but lines on a plot. Learning curves, loss curves, scaling curves— each a whisper of what's next.
They reveal thresholds before we reach them. They expose diminishing returns. They chart the asymptotes of progress.
And they teach us:
More isn't always better. Better is better—when aligned with the curve.

Scaling laws now guide architecture design:

- How deep should the network be?
- How wide should each layer grow?
- How long should you train before gains evaporate?

These are no longer guesses. They are constraints of the frontier. Every dollar spent. Every token processed. Every parameter tuned— now bends toward the slope of the law.

The machine doesn't just learn. We have learned how it learns.

Chapter 25

Mathematical Beauty – Why Truth Feels Elegant

Beauty in mathematics is not decoration—
it is compression without compromise.
A single line that bends a thousand truths into form.
Not because it charms,
but because it cannot be otherwise.

25.1 Elegance as Compression – Simplicity with Power

What makes a formula beautiful?
It is not the curve of the symbols, nor the ink on the page. It is what they compress.

Beauty in mathematics is not aesthetic alone. It is functional.
A beautiful equation expresses much with little.

$$e^{i\pi} + 1 = 0$$

This is Euler's identity. It unites:

- e – exponential growth,
- i – the square root of -1,
- π – the circle constant,
- 1 – the multiplicative identity,
- 0 – the additive identity.

Five of the most fundamental constants in a single whisper.

Elegance is not reduction. It is compression without loss. A distillation that retains depth.
Just as a great poem says in a line what a book cannot, a great theorem unifies truths once thought separate.

Beauty is structure recognized in unexpected places. It is when chaos reveals order. When noise becomes signal. When a pattern becomes universal.
Beauty is resonance— between form, function, and feeling. It is the moment when understanding clicks into place, not because it was forced, but because it fits. Like gravity in orbit, like rhythm in music, mathematical beauty is not an opinion— it is inevitability made visible. It feels less like invention, and more like revelation. As if the truth was always there, waiting patiently behind a veil of complexity, until a single stroke of insight pulled it into the light.
In that moment, you do not merely see the answer— you feel it echo through the structure of thought itself. A theorem becomes a symphony. A proof becomes a poem. And mathematics, stripped of utility, becomes art.

25.2 Symmetry and Surprise – The Dual Nature of Elegance

Beauty is not bland.
A perfect circle is symmetrical— but so is wallpaper.
True elegance arises when symmetry meets surprise.

Symmetry is the promise: that something endures.
Surprise is the twist: that it endures in a way we didn't

see coming.

A mathematical proof is not beautiful because it is obvious. It is beautiful because it seems impossible until it clicks.
The elegance lies in the final connection— when the path becomes inevitable in retrospect.

Fourier's insight: Every sound, no matter how jagged, can be rewritten as pure sine waves.That surprise reshaped physics, engineering, and art.**Gödel's insight:** Within formal systems, there are true statements that cannot be proven within the system. That paradox didn't break mathematics. It made it more honest— and more beautiful. **Elegant mathematics is layered:** You begin with expectation, you encounter tension, and then— you reach a resolution that is simpler than what you thought was possible. **Symmetry is coherence.** Surprise is transformation. Together, they create mathematics that feels alive.

25.3 Aesthetic Selection – Why Mathematicians Trust Beauty

Mathematics is supposed to be cold. Rigid. Unbiased. And yet, ask any mathematician why they believe a result is right— and they may say: *Because it's beautiful.* **Beauty becomes compass.** It points toward what feels correct before the proof is complete. When multiple paths are possible, mathematicians tend to

follow the one that simplifies, unifies, clarifies.
This is not arbitrary. It is pattern recognition honed by centuries of discovery.

Why does this work?

Because beauty in mathematics is often a signal of deep structural alignment.Theorems that are beautiful often turn out to be generalizable. Methods that are elegant often extend beyond their origin.**Beauty is often robustness in disguise.**The physicist Paul Dirac said:

> "It is more important to have beauty in one's equations than to have them fit experiment."

He believed nature favors the elegant. And he was right—his beautiful equation predicted antimatter.This is not mysticism. It is selection pressure.Beautiful theories survive because they compress, explain, and extend.Ugly theories fracture.**Mathematics is not led by taste.** It is guided by resonance.Beauty is resonance—between form, function, and feeling.It is the moment when understanding clicks into place, not because it was forced, but because it fits. Like gravity in orbit, like rhythm in music, mathematical beauty is not an opinion— it is inevitability made visible. It feels less like invention, and more like revelation. As if the truth was always there, waiting patiently behind a veil of complexity, until a single stroke of insight pulled it into the light. In that moment, you do not merely see the answer— you feel it echo through the structure of thought itself. A theorem becomes a symphony. A proof becomes a poem. And mathematics, stripped of utility, becomes art. **Aesthetic Selection** –

Why Mathematicians Trust Beauty

Mathematics is supposed to be cold. Rigid. Unbiased. And yet, ask any mathematician why they believe a result is right— and they may say:

Because it's beautiful. Beauty becomes compass. It points toward what feels correct before the proof is complete. When multiple paths are possible, mathematicians tend to follow the one that simplifies, unifies, clarifies.This is not arbitrary. It is pattern recognition honed by centuries of discovery.Why does this work? Because beauty in mathematics is often a signal of deep structural alignment. Theorems that are beautiful often turn out to be generalizable. Methods that are elegant often extend beyond their origin. Beauty is often robustness in disguise.The physicist Paul Dirac said: *"It is more important to have beauty in one's equations than to have them fit experiment."*

He believed nature favors the elegant. And he was right— his beautiful equation predicted antimatter.This is not mysticism. It is selection pressure. Beautiful theories survive because they compress, explain, and extend. Ugly theories fracture.Mathematics is not led by taste. It is guided by resonance.

25.4 When Beauty Misleads – The Limits of Elegance

Beauty is a compass— but it is not a guarantee.
Sometimes, the most elegant ideas are wrong.

Ptolemy's epicycles were once the height of elegance— nested circles predicting planetary motion with geometric harmony.
But they were wrong.
The heliocentric model was messier— until Kepler's ellipses restored a deeper elegance.

The danger: Mistaking symmetry for truth.
Nature often breaks symmetry to create complexity.
Our desire for simplicity can blind us to reality. **In mathematics:** An elegant conjecture may be false. A simple proof may be incomplete. Sometimes, the truth is inelegant. Sometimes, structure is buried beneath chaos. **Trusting beauty is a heuristic,** not a law.It can guide exploration— but it cannot replace proof. Even in science, some elegant theories (like supersymmetry) have yet to be confirmed. Their beauty seduced generations. Their failure reminds us: nature does not answer to taste. **Mathematical beauty is a signal, not a verdict.** It should evoke wonder, not dogma. To seek elegance is noble. To demand it is naïve. Elegance is not owed to us. It is a gift when it arrives— but a deception when forced.
History is littered with beautiful lies: equations too perfect to be wrong, models too symmetric to doubt,

proofs too clever to question.

And yet, when pressed by experiment, when confronted by contradiction, they crumbled—revealing the cracks beneath their polished surface.

The pursuit of beauty must be tethered to humility. For even the most dazzling structure may be scaffolding around error.

Mathematics does not owe us elegance. Reality does not owe us symmetry. And truth does not owe us comfort.

What remains is the search. And the courage to follow it even when the path twists, even when the shape bends, even when the proof resists elegance.

In that refusal, there is another kind of beauty— the kind born not from harmony, but from honesty.

Mathematical beauty, at its best, is a compass. At its worst, a mirage. It is not the destination. It is the question— echoing through every line, whispering: *Could this be all there is, or is there something deeper still?*

25.5 Beauty as Insight – A Human Need for Coherence

Why does beauty feel like truth?
Because beauty is coherence felt before it is proven.

Our minds are predictive engines— always compressing, modeling, anticipating.
When an idea fits too well, when parts align without force, when many complexities fall into place through a single move—
—we feel something.
That feeling is not proof. But it is real.

Beauty is cognition aligned with compression.
A beautiful proof is not just correct. It makes sense in a way that feels inevitable.
It clicks. It teaches.

This is insight.
And beauty, in its best form, delivers it.
Not through adornment, but through structure. Not through decoration, but through clarity.

Mathematical beauty is not separate from truth.
It is a symptom of it.
When our mental models resonate with the structures beneath the surface of things— we feel elegance not as pleasure, but as alignment.

To seek beauty in mathematics is not to seek distraction. It is to seek understanding with minimal

friction.
To understand something deeply is to see why it could not have been any other way.
That is the moment when truth becomes beautiful. Not because we imposed form upon it— but because the form was already there, waiting to be uncovered, waiting to be seen.
Beauty, then, is not the opposite of rigor. It is its reflection— the echo of structure in the mind that perceives it.
We do not chase elegance to be pleased. We chase it to be aligned.
For in those moments—when the answer feels not invented, but inevitable— we glimpse the architecture of the possible.
And in that glimpse, mathematics ceases to be numbers and rules. It becomes something else:
A language that shapes thought. A mirror that shows the universe as pattern. A flame that lights the way to what we almost knew.

Beauty is not a conclusion. It is the shape of understanding, before we have words for it.

Chapter 26

Mathematical Creativity – Invention Within Constraint

"The artist dreams of freedom; the mathematician thrives on boundaries. For it is within tight walls that the mind finds the narrow path to infinity."
— *Anonymous Principle of Constraint*

26.1 The Paradox of Constraint – Why Rules Ignite Creativity

Creativity is often misunderstood.
We imagine freedom— blank pages, open skies, no rules.
But in mathematics, it is the opposite.
Constraint is the fuel.

A mathematician does not create in a void. They create within systems— axioms, rules, operations.
And precisely because there are limits, they find ways to move, to combine, to leap.

Consider chess. The rules are fixed. The pieces are few.
Yet within that space lie billions of possibilities. Depth. Beauty. Surprise.
Mathematics is deeper still.

Creativity in math is the act of discovering what the rules make possible.
It is about *seeing* what is permitted— but not obvious.
A new proof. A new construction. A new connection between old truths.

Freedom in math is not the absence of structure. It is mastery of it.
When the rules are known well enough, they become instruments.
And like music, the mathematician composes.

Constraint sharpens the mind. Structure invites invention.
To be creative in mathematics is to find surprise in inevitability.
To be creative in mathematics is to find surprise in inevitability.
It is to feel the thrill of a discovery that could not have gone any other way— yet no one had seen it until now.
What seems like imagination is often deep pattern recognition.
What feels like freedom is finely tuned constraint excitability—
a mind attuned to the edges of the possible, dancing at the boundary of proof.

26.2 Invention vs. Discovery – What Does It Mean to Create in Math?

Did Pythagoras discover the relationship between triangles— or invent it?
Did Cantor reveal the hierarchy of infinities— or imagine it into being?

This is the paradox at the heart of mathematics: **Is it out there, or in here?**

To discover is to uncover what already exists. To invent is to bring into being what did not.
In math, both feel true.

The natural numbers feel discovered. They are so fundamental we use them to count ourselves.
But imaginary numbers? Negative dimensions? Fractal infinities?
They feel invented— and yet, once created, they obey rules and reveal truths that surprise even their authors.

Mathematics does not fit cleanly into the discover/invent binary.
Instead, it behaves like language:

- We invent the words.
- But once created, the grammar constrains us.
- New combinations become meaningful within the rules we've agreed upon.

Mathematics is invented structure that enables discovered consequence.

The axioms may be chosen. But their outcomes are not.

To create in mathematics is to build cathedrals on foundations you did not choose— and find that the walls form arches you never imagined.

What begins in abstraction returns with structure.

A guess becomes geometry.

A notation becomes necessity.

In mathematics, the boundary between invention and discovery is not a line—

it is a loop.

We shape the path, but not the destination. And sometimes, the destination teaches us what we meant all along.

26.3 Mathematical Imagination – The Mind as a Laboratory

Mathematics is not written only on paper.
It lives in the mind— a space where logic and imagination cooperate.

Mathematical imagination is not fantasy. It is disciplined vision.
It's the ability to see structures before they're proven. To picture abstract spaces without ever drawing them. Gauss saw surfaces curve in higher dimensions before the tools to formalize them existed.
Ramanujan claimed his results were whispered to him by a goddess— but what he meant was that they appeared fully formed in the landscape of his mind. **This is the work of mathematical creativity:** Not to escape rules, but to feel their implications before they are made explicit.
It is what allows:

- A geometer to see how a manifold folds into itself.
- A number theorist to intuit that a sequence hides a symmetry.
- A category theorist to predict that two entirely different domains are shadows of a deeper correspondence.

Imagination is simulation. It runs mental models before formal construction. The mind becomes a space where mathematics is not only derived— but visualized, tested, and felt. Proof comes later. **Imagination is where math begins.** Proof is where it's made real. But even proof follows the trace of something seen first within. The mind sketches before the pencil moves. What is drawn on paper was rehearsed in silence— a choreography of ideas too fast to write down, yet clear enough to trust. Mathematical imagination is not vague. It is precision before language, structure before symbol, truth before articulation. It is the scaffolding of insight, where future theorems wait to be built.

Some see numbers. Others feel form. But all who imagine in mathematics navigate a realm more internal than seen— a place where the next line of logic is not summoned, but arrives.

This is not mysticism. It is method tuned to intuition. The brain, shaped by structure, becomes a sensor of it. Patterns emerge not by magic, but by resonance with the constraints already held inside. You do not choose to imagine what fits. You follow what clicks. The idea precedes the explanation. The vision precedes the vocabulary.

And when it finally reaches the page— what once lived as intuition becomes form, function, and proof.

Mathematics begins in imagination— but it survives only when the dream holds shape under the weight of logic.

26.4 Generative Boundaries – Why Limits Multiply Ideas

A line on a page is not a prison. It's a prompt.
A rule in a system is not a wall. It's a direction.

In mathematics, the most fertile creativity often comes from **working within strict limits**.

Example: What if you can only move forward and double or subtract 3? How many numbers can you reach? Suddenly, you've entered number theory.
Example: What if you must cover a chessboard with L-shaped tiles— but one square is missing?
You've entered the world of recursive geometry.

Limits sharpen invention.
They force you to recombine. To reimagine. To reverse-engineer.

In combinatorics, the question isn't: "Can you count these?"
It's: "Given these constraints— how many configurations are possible?"

In topology: "What shapes can exist if stretching is allowed, but tearing is not?"

Generative boundaries turn logic into art.
You don't erase the rules. You press into them.
And like clay, they respond with form.

Creativity in math is not just a product of curiosity— but of resistance.

It is the pressure of boundary that gives birth to pattern. And the presence of pattern is not confinement— but the signal that something new is forming.

Mathematics does not sprawl aimlessly. It folds, loops, and spirals within its own limits— like DNA encoding life from four letters, like music unfolding symphonies from twelve notes.

Boundaries are not where creativity ends. They are where it concentrates. Where vague ideas take shape. Where the formless becomes function.

The constraint is not a cage. It is the sculptor's chisel.

A mathematician does not wish for more freedom— but for sharper form. Fewer moves, clearer paths, stronger results.

And in this paradox, creativity thrives: When you cannot do everything, you discover what is truly possible.

26.5 The Infinite Playground

What other field lets you define a world from five axioms— and then explore it forever? Mathematics is not a catalog. It is not a dusty ledger of known results. It is an engine for invention. A playground with infinite terrain.
Every definition is a doorway. Every constraint a corridor. Every new idea extends the landscape in ways no one could predict.

When a mathematician creates, they are not just solving puzzles. They are designing the rules for entire realms.
They define what counts as a move. What counts as a proof. What kind of symmetry is allowed.

Some build new number systems. Some reshape space. Some ask what happens when logic itself is allowed to bend.

In every case— **mathematics is a generative act.**
It is as creative as poetry. As expressive as painting. As infinite as thought. And it is not just the domain of the genius.It is a human birthright— to play with structure, to seek coherence, to build what has never been seen from rules older than language. **Mathematics is not cold. It burns.**It burns with the fire of possibility.

Chapter 27

Mathematical Frontiers

"Mathematics advances not by certainty, but by rupture.
At its frontier, logic trembles, definitions blur, and the unknown demands new language to be seen."

27.1 The Edge of the Known – Where Problems Refuse to Break

Every field has its edges. But in mathematics, the edge is not just a boundary.
It is a pressure point.
A place where thought is stretched to its limit— and sometimes, snaps into something new.

The frontier of mathematics is not a place of confusion. It is a place of sharp clarity where the known meets the unknowable.

These are not unsolved problems because they are vague. They are unsolved because they are precise.
Examples:

- Does $\mathsf{P} = \mathsf{NP}$?
- Are there infinitely many twin primes?
- What is the nature of the Navier-Stokes solutions?
- Is there a pattern to the distribution of the Riemann zeta zeros?

Each problem is a gate— and the key may not be built yet.

Mathematics progresses not just by solving, but by redefining what needs to be solved.
The edge of math is where definitions fail to hold.
Where intuition starts to mislead.

Where new languages must be invented just to describe what we're trying to ask.

At the frontier, math is not about answers. It is about better questions. Questions that reshape the terrain just by being asked. Questions that reveal what assumptions have gone unnoticed.That force us to notice what we thought was obvious.That shift the spotlight from proof to possibility.At the edge, even silence is data. Even failure is movement.
Here, conjectures
become mirrors—showing us how thought behaves when pressed against what it cannot yet grasp.The unknown is not empty. It is structured ignorance. A scaffolding of paradox and elegance, waiting to collapse—or crystallize—under a new formulation. Progress, then, is not always linear. Sometimes, it leaps sideways—inventing whole branches of mathematics in response to a
single stubborn gap. What separates the frontier from the interior is not difficulty— but generativity.The frontier provokes. It invites the mind not only to solve, but to reimagine what it means for something to be solved at all.

27.2 The Unprovable – Gödel's Shadow

In 1931, Kurt Gödel proved something that shook mathematics to its roots.
Not by contradiction. Not by paradox. But by precision.

His Incompleteness Theorems said:

> Even in the most rigorous system, there are true statements that can never be proven within it.

Mathematics had believed itself complete. Perfect. Whole. But Gödel showed that logic has a horizon— and beyond it, truths live without passports.

The First Theorem: Any consistent system powerful enough to describe arithmetic cannot be both complete and consistent.
The Second: No such system can prove its own consistency.

This wasn't a glitch. It was a law.
Truth was deeper than **provability**.

Mathematics now lives under Gödel's shadow.
Every system has limits. Every proof has a boundary. Every axiom carries within it the potential for unprovable truths.

And yet— this is not despair. It is depth.
Because at its edge, mathematics begins to resemble what it had always approached: mystery, not as chaos, but as

structure we are not yet built to reach. —structure
we are not yet built to reach.
A kind of architecture more ancient than proof.
Too complete to collapse. Too intricate to finish.
It is not less real for being unreachable—
it is more real for enduring without us.

We orbit these truths like moons around a dark planet,
pulled by a gravity we cannot land upon.
Each theorem proven is a light lit on the shore,
but the ocean beyond remains infinite.
Not because we failed—
but because mathematics does not end where we end.
It extends beyond certainty,
not as speculation—
but as structure still unfolding.

Every system reveals its edge by working.
Every proof sharpens the outline of what lies beyond.

What we cannot prove
still shapes what we can.
And so, even in silence,
the unprovable speaks.

27.3 Mathematical Intuition – Feeling What We Cannot Prove

Before every proof there is a sense.
A tension in the mind. A resonance in an idea. A glimpse before the light.

Mathematical intuition is not mysticism. It is pattern recognition without formalization.
It is the mind noticing structure before it has language.

You feel it when something seems "obviously true" but resists proof.
You feel it when every example agrees— but the general case stays just out of reach.

Mathematicians speak of beautiful ideas they "know" must be true long before they can demonstrate them. **This is not faith.** It is trained instinct— built from immersion in structure.Like a musician who hears when a note is off even without checking the score.
Intuition guides exploration. It's how Ramanujan saw infinite identities. How Euler leapt from analysis to topology. How Grothendieck conjured spaces so abstract they
required new mathematical senses. **At the frontier,** the proofs come later. The feeling comes first. Mathematical intuition is not a weakness. It is the compass you use when the map ends. And yet—
even without a map,
some minds still move forward. They don't see the path—

they feel the terrain.
Each pattern becomes a foothold,
each tension a clue.
Intuition does not replace rigor.
But it reveals where rigor should look.
It is the murmur beneath the proof,
the ghost of a structure not yet born. It is the murmur beneath the proof,
the ghost of a structure not yet born. Where logic halts, intuition steps forward—
not as certainty,
but as direction. It does not claim "this is the answer," but whispers "look here." In this way,
mathematical intuition is the scout of discovery—
the nervous system of thought,
firing signals before the body of proof can follow. And sometimes,
those whispers become revolutions.
A hunch becomes a hypothesis.
A pattern glimpsed becomes a theorem proved.
Without intuition,
mathematics would be blind to its own edge.
But with it—
it sees just far enough to keep going.

27.4 Future Fields – What Mathematics Doesn't Know It Needs Yet

Some frontiers are visible.
Others are waiting to be named.

Every era of mathematics thought it had seen the full sky— until someone invented a new telescope.
Topology didn't exist until space was allowed to stretch. Category theory didn't exist until structure itself was seen as the object of study.

The future of mathematics isn't just in answering old questions.
It's in discovering new forms of asking.

What new objects can we define?
What operations can we imagine?
What correspondences hide in plain sight between seemingly separate domains? **Artificial intelligence** now suggests proofs humans never would have guessed. **Quantum computation** forces logic to coexist with indeterminacy. **Neuroscience** may one day show why certain mathematical structures feel elegant— and others do not. And still, there will be truths that resist all machines. Still, there will be insights only the human pattern-seeker will feel first. **The frontiers of math are not the edges of the field.** They are the future of thought
itself. They are not just boundaries—
they are beginnings disguised as ends.

The equations we have today
may one day seem like sketches
of deeper laws we have yet to imagine.
What if the next branch of mathematics
is not built on number or space,
but on cognition itself?
What if the axioms of tomorrow
are drawn not from geometry,
but from networks, entropy,
or the architecture of attention?
In every era,
math has mistaken its current frame
for the full frame.
But it is not done expanding.
It never is.
Because mathematics is not a closed cathedral.
It is a living language,
rewriting its own grammar
as it learns to describe
what has not yet been seen.

27.5 The Invitation – What the Unknown Asks of Us

At the edge of the known, there are no instructions. There are no guarantees. No tidy checklists. No promises of completion.

But there is an invitation.
To explore. To ask. To play.

Mathematics does not end. It does not run out.
It waits.
It waits for new questions. New eyes. New frames of reference.

It waits for you.

You need not be a genius. You need not have credentials. You need only to notice a pattern others have not seen—and ask what it might become.

The frontier of mathematics is not protected. It is porous. It lets in those who are willing to follow curiosity past certainty.

The greatest discoveries often begin as a strange feeling that something doesn't quite fit.
That a proof is too clean. That a structure might extend. That a definition might be hiding something wilder underneath.

And in that moment, you're no longer reading mathematics.
You're writing it. You are not filling in blanks.
You are bending the frame.

You are not solving for x.
You are deciding what x should mean.

To write mathematics is not only to reason.
It is to risk.To walk into questions that might not have answers.
To test tools that have never been forged.
To play with objects no one else yet believes are real.And often—
you will fail.You will build castles on sand,
prove the obvious,
chase ghosts.You will write pages that collapse under their own elegance.
You will think you've arrived,
only to find you've circled back to where you began.But this is not wasted motion.This is how mathematics grows—
not as a ladder, but as a branching forest.
Every wrong turn expands the map.
Every dead end teaches what the shape is not.
Somewhere out there,
a twelve-year-old is asking a question no textbook predicted.Somewhere, a janitor in a university is filling notebooks with theorems,
not for publication—
but because the patterns will not leave him alone.Mathematics does not belong to mathematicians.
It belongs to minds that cannot stop wondering. You don't need permission.
You need persistence.
You don't need brilliance.
You need belief—in the unfinishedness of it all. What

matters is not what you know.
What
matters is what you see. A strange loop in a
sequence.
An oddity in an operation.
A space between patterns no one else has noticed yet.This
is where the new begins. Most will not follow you.
Most will stay with the proven, the safe, the solvable.
But you—
you may choose to walk past the edge,
and trust that a bridge may build itself beneath your steps.

Because that's what the unknown asks of us:
To step into its silence
with only the light we carry.
And to trust that when the world is ready—
what we discover will become the next foundation.

Mathematics is not a monument.
It is a motion.
And it needs more movers.

Chapter 28

The Unity of Mathematics – A Hidden Architecture

"The deepest unity in mathematics is not that all answers are connected—
but that all questions, in disguise, are the same."
— Anonymous Pattern-Seeker

28.1 Bridges Between Worlds – When Fields Collide

To the outsider, mathematics appears divided.
There is algebra. There is geometry. There is logic, analysis, topology, statistics.
Each with its own language. Its own culture. Its own terrain.

But beneath the surface, something astonishing happens.
The fields begin to talk.

Algebraic geometry unites equations with shapes.
Fourier analysis turns wave patterns into sums.
Topology meets computation through persistent homology.
Category theory becomes a translator— a Rosetta Stone between formalisms. These aren't just collaborations. They are revelations. Each domain contains shadows of the others. Each discipline, when seen from the right angle, reveals an echo of the same structures.
Mathematics is not a toolkit. It is a single cathedral— constructed from different wings, different vocabularies, but held together by invisible beams. And once you see those beams, you can never go back.

28.2 Deep Correspondences – When Patterns Mirror Across Domains

Some connections are analogies. Others are identities. When mathematics reveals a deep correspondence, it's not a metaphor. It's an equivalence hidden in plain sight. **Dualities** emerge. A structure in algebra is mirrored by one in geometry. A category of topological spaces has an equivalent category of logical statements. This is not coincidence. It is resonance.**Examples:**

- The Langlands program: An audacious vision linking number theory, representation theory, and geometry.
- Pontryagin duality: A bridge between topological groups and their character groups.
- Homology and cohomology: Two distinct ways of measuring shape— which, in many cases, reflect each other.

Mathematics, when seen globally, is not a collection of facts. It is a network of transformations. **The same idea** reappears across disciplines— mutated in form, but stable in essence. Like a melody played on different instruments. You begin to suspect: There is something deeper than the fields. Something the fields are only perspectives on.

28.3 The Hidden Grammar – Category Theory as Unifier

What grammar is to language, category theory is to mathematics.
It does not replace the fields. It reveals how they're structured.

Categories are not collections of objects. They are systems of relationships.
Functions, mappings, morphisms— all become first-class citizens.

In this view, you don't ask, "What is a number?"
You ask, "How do numbers behave when you move between systems?" **Category theory** lets you zoom out. You no longer need to live inside the forest. You can study the relationships between forests. **Functors** map one category to another, preserving structure. **Natural transformations** show how different mappings relate at a higher level. **Adjoint functors** whisper of symmetry in even the most asymmetric domains. Category theory doesn't just organize. It reveals.It shows that many concepts thought unrelated are reflections in a higher-dimensional mirror. When mathematicians speak of
"unreasonably effective" connections, category theory often explains why. It turns collisions into harmonies. It turns fragments into syntax.It gives mathematics its hidden grammar.

28.4 Toward a Grand Architecture – Mathematics as a Living Organism

From the outside, mathematics looks like a library.Isolated shelves. Neatly sorted volumes. Labeled with precision. But from the inside, it feels alive. It evolves. It grows new limbs. It heals contradictions. Mathematics is not static. It is recursive, reflexive, emergent. **New concepts** spawn new structures. **Old contradictions** birth entirely new branches. The discovery of irrational numbers fractured arithmetic and gave rise to analysis. The failure of Euclid's fifth postulate bloomed into non-Euclidean geometry. **The paradoxes of set theory** birthed logic, model theory, and eventually the axioms of ZFC.

It is not a finished building. It is an organism— each theorem like a new nerve, each connection a new artery.

And through it flows something more than calculation: Curiosity. Structure. Elegance. Necessity.

There is no final theorem.
Only unfolding.
Only architecture that builds itself by exploring its own foundations.

28.5 One Field, Infinite Paths

Mathematics is not many things. It is one thing, seen from infinite directions.

You can arrive through geometry. You can arrive through logic. You can arrive through probability, through algebra, through category.
Each path feels distinct. Each has its own language, its own intuition.

But follow it far enough— and they converge.

You see the same symmetry in number and in space.
You find duality in shape and in meaning.
You find computation where you once studied continuity.
There is no center. There is no edge. There is only depth. And every time a new concept is born, it sends out threads— connecting places we thought separate.
Mathematics is not a subject. It is a structure that reveals itself to those who trace what connects everything else. This is the hidden architecture. This is the single field that has no name. And once you walk it, you carry its pattern into every other world you touch.

Chapter 29

Learning Mathematics – Rebuilding Intuition

"Mathematics is not a talent.
It is a language the brain learns to speak—
when given time to listen."
— A Teacher Who Refused to Rush

29.1 The Myth of Talent – Why Understanding Is Built, Not Given

Most people believe math is a gift.
You're either born with "the mind for it," or you're not.

This is false. Worse than false— it's destructive.
Because it takes something that is buildable and calls it unreachable.

Understanding math is not about memorizing formulas. It is not about speed. It is not about being "good with numbers." **It is about structure.** Structure in your thinking. Structure in how concepts fit together. And structure can be built. Every mathematician you've ever heard of struggled. They got stuck. They failed exams. They couldn't see what others claimed was obvious.But they kept building. Not blindly. Not by willpower alone. But by noticing patterns, seeing errors as messages, and reworking their intuitions until the shape became clear. You don't need to be born a prodigy. You need to be patient with confusion. You need to trust that the fog will lift— not because someone tells you the answer, but because your mind, given time and tools, reconstructs the bridge from first principles.

29.2 From Manipulation to Meaning – The Shift to Structural Thinking

Most students first learn math as manipulation. Move the x. Apply the rule. Plug into the formula. Get the answer. Box it. Move on. But something feels hollow. You can solve problems but not explain them. You pass tests but can't teach it.

This is the boundary between procedural and structural thinking. In procedural thinking, you follow steps. In structural thinking, you see relationships. You ask: Why does this operation behave this way? What does this symbol represent in this context? What changes if I flip the frame? Structural thinkers don't memorize rules. They re-derive them on demand. They don't fear new problems. They look for analogies, for mappings between what they already understand and what now sits before them. This shift is not magic. It happens when learners are given space to play with patterns, to notice behavior, to test boundaries, and to ask not "how do I do it," but "why does this work?"

Structure emerges from curiosity. Understanding emerges from playing with form until the rules feel inevitable.

29.3 The Role of Struggle – Why Confusion Is Productive

Most people treat confusion as a stop sign.A reason to quit. A sign that they're not "wired for math." That they've hit their limit.But confusion is not a wall. It is a doorway.Confusion tells you your current mental model isn't enough.It points to exactly where your assumptions break down.And if you linger there— not with frustration, but with patience— you can begin to rebuild.

The most powerful learning moments often begin with being stuck. When something feels wrong, when your answer contradicts your intuition, when a pattern breaks— that's the moment of deepest growth.Because now, you're not memorizing. You're revising your map of reality.True understanding isn't the ability to repeat what you've heard.It's the ability to feel when something doesn't fit— and to sit inside that misfit until clarity takes shape.

Mathematics is made of struggle. Every theorem, every insight, was born in discomfort.You're not behind when you're confused.You're becoming part of the process that built the field itself.

29.4 Language, Vision, and Play – How Different Minds Learn Mathematics

There is no single way to learn mathematics. Some think in symbols. Others in images. Others in stories, rhythms, or motion.**A proof can be a poem.** A function, a melody. An equation, a spatial dance.The mind is not uniform. It is plural.Mathematics itself reflects this plurality. The same idea— like a derivative, or a group, or an integral— can be expressed:

- Algebraically (symbols),
- Geometrically (graphs),
- Verbally (explanations),
- Conceptually (intuitions),
- Physically (manipulatives),
- Visually (diagrams).

And each lens builds a deeper, richer grasp.
Great mathematicians often toggle between lenses. They speak
symbolically, but think spatially. They see algebra as motion. They sketch logic in color. **You are allowed to learn your way.**

29.5 The Long Game – Why Mastery Feels Slow

In an age of speed, mathematics insists on depth.
And depth cannot be rushed.

You will forget. You will relearn. You will feel like you understood something only to lose it the next day.
This is not failure.
This is how long-term structure builds inside the brain.

True mastery is not linear.
It's recursive. You return to old ideas with new eyes.
You spiral upward— each loop deeper than the last.

It takes time for intuitions to form. For symbols to acquire meaning. For patterns to compress. For relationships to light up without conscious effort. This process cannot be shortcut. Fast learners may memorize. But slow learners often see deeper structure. Because they've had to wrestle. They've had to rebuild. They've had to make sense in their own terms. **If it feels slow— you're doing it right.** Mathematics is not a race. It is a cultivation.And the roots that take longest to grow anchor the strongest trees.

29.6 Reconstructing the World

Mathematics doesn't just stay in the classroom. It rewires how you see everything else.

You begin to notice structure beneath the surface.
You spot patterns in nature. You analyze the logic of arguments. You question assumptions with precision.
Mathematical thinking teaches you how to think about thinking. It sharpens your ability to abstract the essential and ignore the noise. It teaches you to model. To simulate. To estimate. To doubt numbers that don't add up and defend ones that do. You see change as calculus. Uncertainty as probability. Structure as logic.
Math is not numbers. It is how complexity is made navigable. It gives you the tools to compress the chaotic into something coherent— without oversimplifying the truth. This is why learning math is not about passing a test. It's about building a mind that can rebuild the world.
A mind that recognizes where the pieces fit, and when they don't.
A mind that doesn't just calculate— but understands.

When you learn mathematics deeply, you don't just learn facts.
You become a builder of clarity.

Chapter 30

Teaching Mathematics – Guiding Minds

"The best teachers do not hand you answers.
They hand you a torch,
and walk with you until you can see the path yourself."
— From a Classroom That Lit the Dark

30.1 The Role of the Guide – Not a Funnel, But a Compass

To teach mathematics is not to fill a vessel, but to orient a mind. You are not a funnel. You are a compass. Teaching is not about transmission. It is about transformation. **The job is not to push information in.** The job is to help students form internal structures that make the information meaningful. This requires restraint.Resisting the urge to solve the problem for them. Resisting the urge to give the shortcut. Resisting the urge to protect them from struggle.Because struggle is how they build. You are not there to make math easy. You are there to make it possible.**You hold the frame, not the solution.** A good guide:

- Asks the right question at the right time.
- Models patient confusion.
- Highlights patterns without naming them.
- Validates effort over correctness.
- Makes space for many ways of seeing.

The best guides know that a moment of clarity earned through wrestling is worth more than a hundred given answers.Because that moment rewires the learner.And that wiring cannot be transferred.It must be grown.

30.2 Safe to Struggle – Creating a Classroom Without Shame

Struggle is not the enemy of learning. Shame is.
You can't think clearly when you feel stupid. You can't take risks when you fear ridicule. You can't explore when you're afraid of being wrong.

To teach math well is to build psychological safety.
Not by removing difficulty— but by removing humiliation.

The classroom must become a place where confusion is normal, not punished.
Where silence is space to think, not proof of ignorance.
Where wrong answers are celebrated as signs of engagement, not signs of failure.

Because the brain only learns when it feels safe.
This is not softness. It's structure.
Safety enables risk. Risk enables insight.

How do you build this?

- Respond to errors with curiosity, not correction.
- Model your own moments of not knowing.
- Use anonymous responses to encourage voice.
- Frame hard problems as opportunities, not traps.
- Replace "What's the answer?" with "What do you notice?"

When students feel that their mind will not be broken for being wrong, they will begin to think more deeply than they thought possible.

The goal is not to prevent discomfort— but to ensure it leads forward, not inward.. Because inward leads to silence. To disengagement. To the quiet collapse of confidence before understanding has a chance to grow. A good math class is not painless. But it is never punishing. It turns mistakes into motion— and fear into fuel. The mind opens not when it is told, but when it feels safe enough to wonder aloud. To risk being wrong in front of others. To test an idea still half-formed. To stumble without retreating. True learning requires vulnerability. And vulnerability requires trust. This is the invisible

architecture of a great classroom: Not rows of desks. Not rigid rules. But a shared agreement— that here, it is safe to not yet understand.

That math is not a judgment. It is a journey.

30.3 Multiple Representations

No two learners see mathematics the same way.
One thinks in pictures. Another in motion. Another in metaphor. Another in sound.

Yet most math instruction is delivered in a single mode—symbols on a board, spoken aloud once.
But math is not a language. It is a world.
And every student needs a way in.

To teach mathematics well, you must teach it in more than one form.

- Draw it.
- Animate it.
- Act it out.
- Map it to a story.
- Ground it in the real world.
- Let students build physical models.

Every new lens unlocks a new mind.
Not because the content changed— but because now it aligns with how they think.

This requires humility.
Your way is not the only way. Your metaphors may fail others. Your symbols may confuse. That is not their fault. It is your opportunity.

The best teachers become translators.

They don't just repeat louder. They switch languages. Until something finally clicks. And a door opens in a mind that had never been reached before. —before the bridge is built between how they see and what the math reveals.

Because understanding isn't delivered. It's assembled. Piece by piece, lens by lens, until the structure becomes visible from the inside out.

Each representation is a tool. Each one reshapes the terrain. Some students need to feel the slope. Others need to hear the rhythm. Still others won't step forward until the equation wears a story they recognize.

This is not decoration. It is design.

Mathematics is not one shape. It is a shape-shifter. The more ways we show it, the more minds we let in.

30.4 The Arc of Understanding – How Concepts Build Over Time

Understanding does not happen all at once.
It builds like sediment— layer by layer, each pass deepening the structure beneath.

To teach well is to honor that arc.
Not to expect mastery on first exposure. Not to confuse temporary performance with deep comprehension.
Understanding begins shallow, and grows through return.

A student might memorize the formula. Then months later, see how it connects to a graph. Then, a year later, realize its real-world application. Then, finally, be able to teach it.
Each loop through the idea tightens the spiral.

As a teacher, your job is to:

- Lay down strong first impressions.
- Return to core ideas in new contexts.
- Connect abstract symbols to lived meaning.
- Allow students to revisit—not just review.

Math is not a list of topics. It is a web of ideas that keep reappearing in new forms.

When students begin to notice those reappearances, when they say, "Wait—that's like what we saw in...," they are not just remembering.
They are integrating.

True learning is not a line.
It's a helix.
You spiral upward by meeting the same idea with more maturity each time. Time is not the enemy of understanding. It is the architecture that holds it together.
We do not learn math by snapshot. We learn it by rhythm— encounter, return, reframe, connect.
The best teaching respects this rhythm. It spaces the moments, not to delay, but to deepen.
Because insight is not a flash. It is a slow accumulation until the weight of connections causes the idea to settle—not on the page, but in the student. Understanding does not arrive fully grown. It germinates. It revisits. It roots itself in time.

30.5 From Fear to Curiosity – Making Struggle Feel Like Discovery

Most students approach math with caution— as if every question is a trap.
They've learned that not knowing is dangerous. That mistakes are marks of failure. That math is a place where you get hurt.
Your job is to undo that.

Because curiosity and fear cannot share the same space.
Fear shuts down exploration. Curiosity opens it.
And mathematics is built on exploration.

So how do you shift fear into curiosity?
You reframe the unknown as an invitation.
You say, "I don't know either—let's find out."
You reward insight that comes after a mistake, not just correctness on the first try.
You model the joy of wrestling.
You teach them that struggle is not a sign of weakness— but the beginning of understanding.

When students realize they are allowed to not know, something opens.
They ask bolder questions. They play with ideas. They begin to wonder again.

That's when real learning happens.
Not when they fear the problem. But when they chase it.

The goal is not comfort. The goal is courage. And courage, in math, means staying with the puzzle long enough to let it change you. You don't need to have all the answers. You just need to create the conditions where answers can emerge.

Curiosity doesn't grow in silence. It grows in the open— where ideas are spoken, even if uncertain, especially when uncertain.

Let students hear the questions you still carry. Let them see that the journey never ends— it only deepens.

Because mathematics is not a ladder. It's a labyrinth.

And when they learn that every wrong turn is a new hallway, they stop fearing the maze.

They start exploring it.

That shift— from evasion to engagement, from trembling to tinkering— is the transformation that opens the mathematician within.

30.6 The Teacher's Growth – Becoming a Better Guide by Letting Go

To teach mathematics is to evolve with it.
Not just the content. The craft.
Because the more you teach, the more you realize: you are not the hero of their learning journey.
You are the terrain they walk through. The mirror they bounce their thinking off.

Your authority is not in having the answers— but in building the environment where answers can emerge.

That means letting go of:

- Needing to be right.
- Rushing to correct.
- Over-explaining.
- Filling every silence.
- Teaching the way you were taught.

And embracing:

- Asking better questions.
- Designing better problems.
- Noticing student thinking with compassion.
- Accepting ambiguity as part of the path.
- Learning from every moment that didn't go well.

Growth as a teacher is not a steady climb. It's a humbling spiral.
You return again and again to the same core truth: teaching is not about control.
It's about trust.
Trust in the learner. Trust in the process. Trust in yourself— not as the deliverer of math, but as the architect of understanding.

You don't need to carry them. You just need to clear the path, light it well, and walk alongside until they can navigate it on their own.

Chapter 31

Mathematics and Meaning – Why It Matters Beyond the Classroom

"Mathematics, rightly viewed, possesses not only truth, but supreme beauty." — Bertrand Russell

31.1 More Than a Subject – Math as a Way of Seeing

To most, mathematics is a school subject. A set of procedures. A sequence of hurdles. A gatekeeper to diplomas.

But mathematics was never meant to live only in classrooms.

It is not confined to tests. It is not made of tricks.

Mathematics is a way of seeing.

It sharpens perception. It clarifies structure. It reveals connections others might miss.

A child who sees number patterns begins to sense rhythm in chaos.

A teen who understands scale begins to grasp the power of exponential change.

A citizen who understands statistics can see through false claims with confidence, not cynicism.

Math gives us more than formulas. It gives us lenses.

To teach math is to hand students tools for decoding reality.

And once these tools are internalized— they don't disappear after graduation.

They stay with you when you read the news. When you weigh a risk. When you notice an injustice that rests on bad numbers. **Math becomes not just knowledge— but defense.**

Against manipulation. Against confusion. Against the lie that you must trust what others say is true.

And even deeper: math becomes a source of meaning.

Because when you see how the pieces fit, how the world folds together, even for a moment—

you glimpse something sublime.

In that glimpse, you feel it: That behind the noise, there is form. That beneath the arbitrary, there is order. Not imposed from above— but arising from within.

Mathematics offers not just answers, but the quiet astonishment that comes when something aligns. It reminds you that beauty isn't decoration. It's the resonance of parts that fit. Of patterns that persist. Of truths that do not change when you look away.

In this sense, mathematics is not cold. It is clarity made intimate. It is reason lit with wonder. It is the architecture of meaning that doesn't collapse when feelings shift. To see mathematically is not to strip life of magic— but to realize the magic was in the structure all along.

31.2 Math as Empowerment – Building Agency in an Engineered World

You live in a world shaped by mathematics.
Your city, your phone, your job applications, your medical scans, your voting systems—
all run on algorithms, data models, probability curves, cost functions, feedback loops.

To understand math is to understand the levers of the modern world.
And to not understand it— is to be acted upon without defense.

Mathematics is not a luxury. It is not just for the "gifted."
It is a tool of agency.
It lets you:

- Spot manipulation in a misleading graph.
- Challenge the fairness of an algorithm.
- Decode the logic behind a system's behavior.
- Model possible futures with clarity and calm.
- Invent solutions no one else could see.

Math is not obedience to rules. It is the power to shape rules. To bend systems toward equity, clarity, sustainability.

When students realize this— that math is not just a hoop to jump through, but a door to influence— everything shifts. They begin to engage not for the grade, but for the power. Not for compliance, but for comprehension. **Math becomes language. Math becomes leverage.** And in a world increasingly driven by hidden calculations, that might be the most important literacy of all. Because when you understand the structure, you are no longer a subject of the system— you are a participant in its design.

Mathematics gives you more than comprehension.

It gives you voice.

Not just to speak up, but to model why something matters, to prove why something should change, to architect a better alternative. This is not about turning everyone into a mathematician. It's about ensuring no one is left powerless in a world built from math-shaped decisions. Empowerment begins not with knowing all the answers, but with knowing how to ask the right questions— and how to follow their consequences with rigor and clarity.Math, then, is not just knowledge. It is ethical infrastructure. It is how you claim your place in a world written in patterns.

31.3 The Moral Frame – Numbers Are Not Neutral

We are told that numbers don't lie. That mathematics is objective. That it is immune to bias, to agenda, to politics.
But this is only half true.

Yes, numbers follow rules. Equations do not favor sides. A triangle's angles add to 180° whether drawn by emperor or exile.
But the ways we use numbers— the systems we build with them— are deeply human.
And humans are never neutral.

Every dataset reflects choices:

- What do we measure?
- What do we ignore?
- Who collected the data?
- Who was left out?

Every algorithm embeds assumptions. Every model simplifies a world that resists simplification. **And every equation can be used to reveal truth— or to obscure it.** To teach mathematics without teaching ethics is to hand someone a weapon without teaching when to use it. A predictive policing algorithm is built on probability. But it can deepen injustice if trained on unjust histories. A medical triage model can save lives—

or

reinforce bias. Even something as simple as a test score can shape a future in ways that reflect not talent, but inequality. **Mathematics gives power. But without ethics, power corrupts.** So let us teach math with care. With context. With conscience. Because numbers do not lie— but people can through how they use them. Mathematics, then, is not just a tool of precision— it is a mirror of our values. What we choose to model reveals what we consider important. What we exclude reveals what we are willing to forget.And so, math becomes a kind of ethics in code: a language through which we encode what we care about, what we prioritize, what we make visible. Teaching mathematics must include teaching this mirror. Not just to sharpen minds, but to shape judgment. Because in the hands of the compassionate, a formula can heal. But in the hands of the careless, it can erase.

The responsibility of mathematics is not only to be right. It is to be just.

And that responsibility lives not in the numbers— but in the ones who use them.

31.4 The Search for Meaning – When Math Points Beyond Itself

For some, mathematics is a tool. For others, a game. But for a few— it becomes something else entirely.
A path. A mirror. A portal to meaning.

There is a moment— in every mathematician's life— when a proof feels like revelation.
When a pattern long hunted suddenly locks into place, not as a trick, but as something that had always been true.
And for a second, you are not just solving the problem. You are communing with structure itself.

This is the secret that textbooks often forget: mathematics can be spiritual.
Not in a supernatural sense— but in the awe it evokes. The silence it commands. The sense that truth exists independent of us and we, briefly, touch it.

Mathematics points beyond itself— to the hidden architecture of reality.

- The primes whisper of infinity's fingerprint.
- The Mandelbrot set reveals endless variation within order.
- Gödel's incompleteness shows that even truth has boundaries it cannot cross.

To students, these ideas need not be abstract. They can be invitations.

Not just to calculate, but to contemplate.

Math, then, becomes not only power—

but poetry.

A language capable of not just describing the world, but making it feel more meaningful. Because when you trace a curve to its limit, or glimpse symmetry where none was expected, you begin to wonder: Is this beauty discovered— or invented?

Mathematics invites that question again and again. Not to answer it, but to keep you circling it with wonder. In the most rigorous proof, there is room for reverence. In the coldest abstraction, there can be warmth— a feeling that this, this shape, this truth, was waiting to be seen. You are not just building arguments. You are uncovering echoes— as if the universe has always spoken this language, and you are learning, at last, how to listen. Mathematics does not require belief. But it offers belonging. A place where minds can meet meaning not through opinion— but through structure. And perhaps that is why, at its highest level, math feels like coming home to a place you never knew you left.

31.5 Mathematics as a Life Practice – Thinking Clearly in a Noisy World

Mathematics does not end when the class bell rings. It is not sealed in exams or confined to chalkboards.
It becomes a habit of thought.

To reason precisely. To seek what's consistent. To question what feels obvious. To simplify without distorting. To build from first principles when noise overwhelms.

In a world flooded with misinformation, this is not a luxury. It is a survival skill.
Mathematics becomes a practice— of clarity.
Like meditation sharpens attention, like music trains the ear, math refines your ability to think.

It does not demand belief. It demands understanding. It does not ask you to memorize. It invites you to reconstruct.

And the more you engage with it, the more you notice its echoes:

- In the fairness of a split check.
- In the probability of a risky choice.
- In the symmetry of a design.
- In the structure of an argument.

You begin to live mathematically— not by turning everything into numbers, but by treating complexity with patience and structure and humility. **You do not worship certainty. You learn to walk with it— without losing balance in the unknown.** That is why mathematics matters. Not just for science. Not just for tech. But for the mind itself. Because in learning math, you practice being clearer, calmer, and more human in how you think. And in that clarity, you begin to navigate the world differently. Not by having all the answers, but by asking better questions. You stop rushing to conclusions. You pause to examine assumptions. You trace implications before reacting. You see error not as failure— but as information. Mathematics becomes a way to carry yourself— with precision, with care, with the quiet strength of someone who knows how to test a claim before accepting it. It teaches you to be rigorous without being rigid. To doubt with purpose. To seek simplicity that is earned, not imposed. And in a time when certainty is often faked, when confidence drowns out correctness, this way of thinking is an act of resistance. Mathematics, then, is not just a subject. It is a stance. A choice to be mindful, methodical, and meaning-driven— even in the noise.

Chapter 32

The Future of Mathematics – Open Questions and New Worlds

"The universe is under no obligation to make sense to us. But mathematics is how we try to listen anyway." – *Unknown*

32.1 The Frontier Is Not Finished — Problems That Still Refuse to Yield

For all its triumphs, mathematics is not complete.
It is not a closed museum of polished ideas. It is a living, breathing endeavor— with edges still jagged, with doors still locked.

There are truths we suspect exist, but cannot prove.
There are patterns we glimpse, but cannot yet explain.
And for every solution, new questions unfurl.

Here are a few of the riddles that remain:

- **The Riemann Hypothesis** — Is there a hidden pattern to the primes deeper than any we've found?
- **P vs NP** — Can every problem that can be verified quickly also be solved quickly?
- **The Continuum Hypothesis** — Is there a size of infinity between the integers and the real numbers?
- **The Navier–Stokes Equations** — Can we predict the motion of fluids under all conditions— or are there states where chaos takes over?
- **The Collatz Conjecture** — Can the simplest rule create infinite complexity from every starting point?

These are not just puzzles. They are mirrors. They show us the limits of what we know.

And in that humility, mathematics grows.

A student might think the map is complete— that math is solved.
But those who stand at the edge of the field know otherwise.
They see dark valleys and distant peaks, and they keep walking.
Because the frontier is not finished.
It is still unfolding.

Some frontiers can be seen. Others must be imagined.
Mathematics is not waiting to be finished. It is waiting to be transformed.
Each unsolved problem is more than a challenge. It is an invitation to evolve thought itself.
To approach these mysteries is not merely to seek answers— but to build better questions. Questions that will one day teach us how to think in dimensions we don't yet know we're missing.
Because these riddles do not just resist us. They reshape us.
The frontier is not where mathematics ends. It's where it begins again— with every mind brave enough to ask what lies beyond.

32.2 New Languages – Inventing Tools to See What We Couldn't Before

Mathematics evolves not just by solving old problems—but by inventing new ways to think.
Sometimes the problem remains unsolved because the language is incomplete.

What calculus did for motion, what set theory did for infinity, what topology did for shape—
new structures are born when old ideas no longer suffice.

Today, new languages are emerging.

- **Category theory** reframes mathematics not in terms of objects, but in terms of relationships.
- **Homotopy type theory** bridges logic and geometry, redefining what it means to prove something.
- **Model theory** asks: what truths can be known within one mathematical universe— and which break when viewed from another?
- **Computational mathematics** introduces symbolic engines and proof assistants— not as shortcuts, but as companions in thought.

Language is not neutral. It guides what can be seen.

When mathematicians shift perspective, when they invent a new grammar— new truths become possible.

A rigid proof can become a living system. A collection of numbers can become a flow. A tangle of ideas can become a map when the right concepts arrive.

The frontiers of math are not just in content— but in the very form of its expression.
Each new formalism is not a cosmetic change. It is a cognitive evolution.
It is the birth of a deeper way to see.

The next revolution may not come from solving harder equations— but from asking them in a different tongue.
Every generation of mathematicians inherits not just problems, but the shape of thought itself. And that shape can be reshaped. A new notation isn't just a symbol. It's a scaffold for ideas not yet possible. A new logic isn't just a system. It's a lantern into regions of structure no mind has walked before.
This is not mathematics becoming more complicated. It is mathematics becoming more aware of itself.
To build a new language of thought is to remake the space in which meaning is formed.
And from that space, whole new worlds may rise.

32.3 The Interdisciplinary Future — Mathematics as Translator

Mathematics has always been more than its own domain. It is the quiet language behind chemistry's bonds, behind physics' particles, behind computer science's logic, and economics' models.

But the 21st century is revealing a new role: mathematics as translator.

In an age of complexity, no single field is enough. Biology speaks in networks. Neuroscience in probability. Linguistics in structure. AI in optimization and .
And across them all— mathematics whispers shared forms.

- Graph theory models the spread of disease and ideas alike.
- Information theory unites neurons, machines, and genomes.
- Topology helps study the shape of data as much as the shape of space.
- Logic and computation underlie everything from proof assistants to legal systems in code.

Mathematics is becoming the connective tissue of thought. It allows physics to speak to biology, statistics to speak to ethics, and machine learning to speak to our most human concerns. In this way, math is not only a tool. It is a grammar for translation— for finding shared structure in systems that seem unrelated. **The future belongs to those who speak many fields—but think mathematically.** Not by memorizing formulas, but by seeing the patterns beneath phenomena. By turning chaos into coherence. and disconnection into dialogue. Mathematics does not erase the uniqueness of each discipline. It honors it—by revealing what can be shared without being simplified. When a concept appears in both neuroscience and network theory, it is not coincidence. It is resonance.

When entropy speaks through both thermodynamics and linguistics, it is not metaphor. It is signal. To think mathematically across domains is to notice what others overlook: That life, language, and logic are all dances of constraint and possibility.That structure repeats—not because the world is simple, but because it is lawful.

In this future, the mathematician is no longer the gatekeeper. They are the bridge builder.

Between data and meaning. Between biology and behavior. Between technology and ethics.

Because when everything is connected, what we need most are those who can see how.

32.4 The Limits of Formalism – What Math May Never Capture

For all its power, mathematics is not omnipotent.
It is a system of symbols— a logic of rules and rigor.
But not everything that matters can be formalized.

There are truths that resist translation into numbers or proofs.
What is the experience of color? What is the meaning of suffering? What is the weight of a memory?
These are not errors of measurement. They are the boundaries of the symbolic.

Gödel taught us that even in pure math, some truths are real but unprovable.
Turing taught us that some answers will never be computed— no matter how fast the machine.
There is structure in the world that cannot be reduced without being destroyed.

This does not make mathematics weak. It makes it honest.
To know its domain is not a limit— it is a virtue.

There will always be tensions between precision and presence, between proof and perception.
And as math evolves, it may reach toward those tensions— but never erase them.
Mathematics is powerful— but it is not the world.

It is a map. A lens. A scaffold for understanding.
But the human experience will always extend beyond what theorems can hold. And that is its beauty— not in pretending to answer everything, but in illuminating what remains unanswered.
Mathematics brings clarity where clarity is possible, and humility where it is not. It does not dissolve mystery. It sharpens our gaze around it. To confront the limits of formalism is not to abandon mathematics— but to use it wisely. To see it as one thread in the fabric of knowing. One language among many. The mathematician who knows where not to calculate is no less rigorous— they are more human. Because the goal is not to turn life into numbers, but to meet life with structure and to leave space for what cannot be structured.
In this way, mathematics becomes not an answer— but a companion in the search.
A way of walking toward meaning without pretending to enclose it.

32.5 Mathematics Will Never Be Finished

There is no final theorem.
No last chapter. No perfect equation that explains it all.
Mathematics is not a monument. It is an organism— growing, fracturing, branching into unknown space.

Each time we think the map is complete, a new pattern appears at the edge of vision.
A contradiction opens a door. A strange coincidence becomes a field. A child's question shatters centuries of structure.

What drives it all?
Not utility. Not certainty.
Curiosity.
The refusal to accept that what we see is all there is.

Mathematics is not just the language of science. It is the poetry of logic. The art of structure. The dream of precision in a world that often refuses it.
And like all art, it is never complete.

As long as there are patterns, as long as there are minds willing to chase them— mathematics will evolve.
Not because it needs to. But because we do.
Because in seeking the infinite, we understand the finite more clearly.
Because in building what has never existed, we become something new.
Mathematics will never be finished— and neither

will we. Because mathematics is not only what we study. It is what we become in the studying.

The patience to sit with ambiguity. The courage to pursue clarity. The wonder that a blank page invites.

It is not just that mathematics grows— it grows us.

Each conjecture is a question we dare to ask. Each proof, a path we forge through the unknown. Each unsolved problem, a mirror held up to our imagination.

And so we return—again and again— not to reach an end, but to remain in motion.

Not to conquer mystery, but to be reshaped by it.

There is no end to this journey. Only thresholds.

Only new ways of seeing, and new selves awakened in the seeing.

Chapter 33

Epilogue – A Map, A Mirror, A Flame

"Mathematics is not the cold machinery of reason. It is the fire we use to see."
— Anonymous

33.1 A Map of Mind Made Visible

This book is not just a tour through the territories of mathematics.
It is a portrait of how thought moves.

Each chapter was a room in a larger architecture.
A system of structures woven from symbol and meaning.
A way of seeing that does not merely describe reality—but composes it.

Mathematics is the mind drawn outward.
A record of every pattern we've dared to touch and name.
It is not a thing we study— but a trail we leave behind as we try to make sense of the world.
It is a map. But not of terrain. Of possibility.

33.2 A Mirror for Being

Mathematics reflects not just what we know— but how we know.
It shows us our desire for structure, our tolerance for ambiguity, our obsession with truth.

In its symmetry, we see our hunger for balance. In its abstraction, we see our need to transcend the immediate. **In its paradoxes, we see the limits of what we can bear.**

Mathematics is not emotionless. It is clarity in pursuit of the essential.
It is our most precise language— and sometimes, our most beautiful one.

It is a mirror. Not of appearance. But of understanding.

33.3 A Flame That Never Dies

Mathematics is alive because questions remain.
Because wonder persists.
Because minds that do not yet exist will pick up this trail and walk further than we could imagine.

We do not preserve mathematics to store knowledge. We keep it burning so that meaning continues to unfold.
Every insight passed on is a spark.
Every child who asks why is a carrier of the flame.

And that flame does not end.
It flickers in proofs and paradoxes. It leaps between geometry and computation. It glows in quiet moments of realization in classrooms and notebooks all over the world.
It is a flame. Not of destruction. But of illumination.
And as long as we chase what is hidden— what lies beyond the next theorem, the next idea, the next understanding—
it will burn.

Appendix: A History of Mathematics – The Long Awakening of Structure

Before the First Equation

Before there was number, there was subtraction.

A hunter sees four deer. He strikes one. Now he sees three.

He doesn't speak it. He doesn't write it. But something inside him senses the loss. The quantity has changed.

This wasn't mathematics. Not yet. It was life felt through *difference.*

Somewhere around 43,000 years ago, a human carved notches into a baboon's fibula—the *Lebombo Bone*—to mark time, moon phases, or maybe menstrual cycles. These were not symbols of mathematics. They were records of *occurrence.* Repetition had rhythm. Rhythm had meaning. And meaning had to be remembered.

In those marks lay the seed of number.

From Crops to Clay

As societies formed, mathematics became necessity.

The Sumerians pressed marks into clay tablets to track harvests, debts, and trades. The *base-60 system* they devised—chosen perhaps for its divisibility—gave rise to our modern measures of time (60 seconds, 60 minutes) and the circle (360 degrees).

To measure a field after the Nile had erased its boundaries, the Egyptians turned to geometry: *geo-metria*—"earth measurement." What began as a practical art would become the earliest *architecture of space.*

Wherever civilization grew, mathematics followed. Not as theory, but as *tool.* It counted grain. It predicted floods. It built pyramids. It served the state, the temple, and the merchant.

But theory was coming.

The Greeks and the Leap to Proof

The Greeks did not invent mathematics.

But they did something unprecedented: they *proved* it.

Where Egypt measured, the Greeks abstracted. Where Babylonians computed, the Greeks *deduced.* Pythagoras didn't just observe the triangle—he saw in it a *harmony of form.*

In *Euclid's Elements*, written around 300 BCE, mathematics became a kind of sacred language—built from definitions, axioms, and logical inference. Proof became the gold standard. Certainty became possible.

Yet even the Greeks stumbled upon paradox. They discovered numbers that couldn't be written as ratios: *irrational numbers*, like the square root of 2. These were mathematical intruders—proofs that something in their system didn't align with their belief in wholeness.

Still, they pressed forward. To the Greeks, mathematics was not a tool. It was *reality's underlying order.*

Zero, Infinity, and the Indian Revolution

While the West still feared the void, Indian mathematicians embraced it.

They invented *zero*—not just as a placeholder, but as a concept. A symbol that stood for *nothing*, and by standing for it, made *everything* computable.

They explored *infinity* not just as a theological abstraction but as a mathematical reality. By the 7th century, *Brahmagupta* had already developed rules for zero and negative numbers—ideas that would take Europe centuries to accept.

And then came the *decimal system*: place value, simplicity, elegance. The Indian system—passed to the Islamic world and eventually into Europe—became the invisible skeleton of modern computation.

Algebra and the Islamic Golden Age

In the 9th century, in the House of Wisdom in Baghdad, *Muhammad ibn Musa al-Khwarizmi* wrote *Al-Kitab al-Mukhtasar fi Hisab al-Jabr wal-Muqabala*—"The Compendious Book on Calculation by Completion and Balancing."

From it came the word *algebra.*

In the Islamic world, mathematics became a bridge: between East and West, between number and meaning, between earth and sky. Scholars preserved Greek knowledge, translated Sanskrit texts, and innovated new systems of their own—all while exploring the mathematics of astronomy, architecture, and inheritance. Where Europe had descended into the shadows, Islamic mathematics kept the flame alive.

The Renaissance: Number Reborn

Then the light returned to Europe.

Fibonacci, traveling through North Africa, brought back the Hindu-Arabic numeral system—and with it, a new efficiency in thought. His *Liber Abaci* introduced a continent still tangled in Roman numerals to the simplicity of 1 through 9, and the silent power of *0.*

The *printing press* amplified mathematics. What once had to be copied by hand could now spread with ink and speed.

Then came *Descartes.* With the Cartesian plane, he united algebra and geometry—drawing equations as shapes, and shapes as equations. Mathematics became not only a tool of the world but a *map of it.*

Calculus and the Language of Change

At the dawn of the Scientific Revolution, *Isaac Newton* and *Gottfried Wilhelm Leibniz*—independently and nearly simultaneously—

discovered *calculus.*

The world no longer needed to be measured in frozen snapshots. It could be understood in *motion.* Rates of change. Accumulations of small pieces. The slope of a tangent. The area under a curve.

Calculus allowed humans to measure not just what *was*, but what *became.*

And it allowed mathematics to speak, for the first time, in the grammar of *change itself.*

The 19th Century: Infinity Cracks the Mirror

The 1800s brought brilliance—and breakdown.

Carl Friedrich Gauss revealed the hidden structure in numbers and space. *Évariste Galois* uncovered the deep symmetries of algebra.

But it was *Georg Cantor* who shattered the old order. He showed that there were different *sizes of infinity.* That the infinite could be mapped, counted, compared. And that some infinities were larger than others.

Meanwhile, *non-Euclidean geometry* emerged from the shadow of the fifth postulate. *Lobachevsky*, *Bolyai*, and *Riemann* imagined geometries where parallel lines diverged or converged—geometries that would, decades later, become the canvas for Einstein's theory of general relativity.

Mathematics had broken free of physical space. It was now a universe unto itself.

The 20th Century: Crisis, Machine, and Mystery

David Hilbert believed mathematics could be made whole. A complete system. A firm foundation. Everything provable from axioms.

Then *Kurt Gödel* shattered that hope.

In 1931, Gödel's *incompleteness theorems* proved that any sufficiently powerful mathematical system would contain truths that could not be proven within it.

Mathematics had limits—not just practically, but *fundamentally.*

And while Gödel looked inward, *Alan Turing* looked outward— imagining a machine that could simulate any algorithm. His *Turing machine* became the blueprint for modern computation.

Then came *chaos theory*: small changes leading to wild outcomes. Then *fractals*: infinite detail emerging from simple rules.

The certainty of earlier centuries gave way to a deeper truth: Mathematics does not control the world. It reveals its *complexity.*

The Present – Maps of Maps, Patterns Without End

Today, mathematics maps its own structure.

Category theory doesn't care what objects are. It cares how they relate. *Topology* ignores distance, focusing instead on connection. *Networks*, *graph theory*, and *information theory* now underpin everything from social media to neuroscience.

In physics, mathematics remains the backbone—from the Standard Model to quantum mechanics, from black hole entropy to string theory. It describes what we cannot see. It predicts what has not yet been observed.

But it also expands into new terrain: AI, data science, bioinformatics, quantum computing. Mathematics is no longer only about *truth*. It is about *pattern*.

Compression. Prediction.

And still—the most ancient question remains: Is mathematics invented, or discovered? Is it the scaffolding of the cosmos? Or a mirror, reflecting the architecture of our own minds?

Maybe it is both. Maybe it always was.

Mathematics as a Timeline

This map was not drawn in a day.
It unfolded, idea by idea, across generations—
Not as a line, but as a pattern
that humanity slowly learned to see.

- **-44,000 BCE** — *Lebombo Bone*, notched to track the moon: cognition externalized through tally.
- **-20,000 BCE** — *Ishango Bone* emerges near the Nile: multiplicities of two and prime marks, still debated.
- **-3000 BCE** — Sumerians shape clay into arithmetic. Base-60 dawns.
- **-2000 BCE** — Plimpton 322 records Pythagorean triples before Pythagoras breathes.
- **-1800 BCE** — Egypt maps slope and fraction on papyrus. Rhind, Moscow: documents of practical geometry.
- **-800 BCE** — The Shulba Sutras construct altars with root-two precision. Ritual becomes ratio.
- **-600 BCE** — Thales sees geometry as structure. Theorems leave the stars and touch earth.
- **-300 BCE** — Euclid births axioms. Proof begins.
- **-260 BCE** — Archimedes carves curves. Proto-calculus takes form.
- **800 CE** — Al-Khwarizmi names algebra. "Al-jabr" writes balance into equations.
- **1200 BCE–300 BCE** — Chinese rod numerals count through contact. Arithmetic held in hand before mind.

- **1637** — Descartes merges space with symbol. Coordinates give thought direction.
- **1687** — Newton and Leibniz fracture motion into moments. Calculus begins to flow.
- **1874** — Cantor opens the doors of infinity. Some sets are larger than others.
- **1905** — Einstein writes space and time into curvature. Reality bends mathematically.
- **1931** — Gödel draws the first internal fracture. No system sees all its own logic.
- **1936** — Turing encodes thought in tape. Machines awaken from structure.
- **1948** — Shannon names information. Bit by bit, communication becomes quantifiable.
- **1976** — Computers co-author proof: the Four Color Theorem is drawn to a close.
- **1994** — Wiles finishes Fermat's whisper across centuries.
- **2010s** — Matrices begin to mimic minds. Neural networks rise.
- **2023** — GPT-4 launches. Billions of weights shaped by gradient descent into generative coherence.
- **2023** — Proof verification becomes algorithmic: Lean, Coq, Isabelle assist human rigor.

- **2023** — Liquid Neural Networks shift toward dynamical systems. Learning becomes fluid.
- **2024** — Symbolic regression rediscovers physics. Theory emerges from raw data.
- **2024** — Category theory reclaims its place—compositional AI begins to model structure, not just surface.
- **2024** — Tensor networks compress complexity. Quantum behavior becomes computable.
- **2025** — Machines offer theorems. Insight is no longer human alone.
- **2025** — Integrated Information Theory simulates consciousness as a geometry of information.
- **2025** — Topology meets transformers: persistent homology interprets neural dimensions.
- **2025** — Analog substrates stretch the Universal Approximation Theorem into new silicon.
- **Ongoing** — Mathematics is not bound to blackboards. It lives in lenses, predictions, the curvature of your GPS.
- **Today** — The map expands not by destination, but by dimension. The page you read is not outside it.

About the Author

"I did not create the shape. I followed it."

Joel Peña Muñoz Jr. writes the way others listen—to something that's already there
He was raised with more questions than instructions, more silence than certainty. The lines he draws across the page are attempts to make sense of the quiet rules beneath everything.
He does not claim to own the knowledge in this book. He merely wandered far enough to find it waiting.
The Shape of All Things is a map he followed, not one he made. A trail of logic, wonder, contradiction, and clarity—left behind by the universe, waiting to be read.

He is grateful for the teachers who didn't give up, the family who let him wander, and the quiet rooms where curiosity was allowed to bloom.

About the Author